Bringing the Mediterranean into Your Garden

How to capture the natural beauty of the garrigue

OLIVIER FILIPPI

Translated by Caroline Harbouri

filbert *press*

Originally published in France as *La Mediterranée dans votre jardin* in 2018
by Actes Sud, Le Méjan, place Nina-Berberova, 13200 Arles, France
Copyright © Actes Sud 2018

English language edition copyright © Filbert Press 2019
All rights reserved

ISBN 978-1-9997345-1-0

All photographs are by the author unless otherwise noted

A catalogue record for this book is available from the British Library

Printed in China

Front cover: The garden of Vilka and George Agouridis, near Athens, where landscape designers Jennifer Gay and Piers Goldson have drawn inspiration from the garrigue to create an evergreen scene where shades of green, grey and silver remain attractive throughout the year.

Back cover top: The mediterranean section of the Prague Botanical Garden with *Cistus laurifolius, Achillea coarctata, Tanacetum densum, Lavandula angustifolia* and *Hypericum olympicum*, as well as different species of thyme, helianthemum, sedum and micromeria.

Back cover bottom: The gravel garden at The Beth Chatto Gardens near Colchester. Gardening on gravel is one of the best ways of growing garrigue plants outside mediterranean-zone climates.

CONTENTS

A NEW SOURCE OF INSPIRATION FOR GARDENS

When you stand on a crag buffeted by the mistral above the calanques of Marseilles, or make your way along a rocky path in the mountains of Greece, or follow the ancient mule track that crosses the Serra de Tramuntana in Mallorca, you may be struck by the extraordinary beauty of the landscape around you. This is the garrigue, a landscape typical of the area surrounding the Mediterranean Basin, which is dominated by shrubs and lower sub-shrubs growing in arid conditions where the stony soil, usually limestone, generally remains partially visible between the plants.

As you look out on vegetation sculpted by the wind or by goats, foliage from dark green to silver set off by the power of the stony environment and brilliant light that emphasizes the rhythm of the ground-hugging cushion shapes of the plants – and you inhale an intensity of scents in which the essential oils of helichrysums and sages are mingled – suddenly you become aware that you are contemplating the combined elements of a magnificent garden, with one difference: there is no gardener here to weed or water, prune the plants or give them fertilizer.

Around the Mediterranean Basin these beautiful garrigues, sometimes covering only a few square metres and elsewhere spreading over an entire hillside as far as the eye can see, form a remarkable mosaic of autonomous 'landscape gardens' that motivate the gardener to start thinking and asking questions. How can one apply the conditions of the Mediterranean garrigue to reduce maintenance in one's own garden? Which of these plants would flourish in such a garden? How should one plan the planting of a garden to evoke the particular beauty of the garrigue?

Garrigue plants have long been an integral part of the human environment for their medicinal properties or domestic uses. According to legend, it was in the hollow stem of a giant fennel, whose spectacular inflorescences light up the landscape in late spring, that Prometheus carried to mankind the fire which he had stolen from the gods on Mount Olympus. In the first century CE the Greek physician Dioscorides drew up a list of the plants used in the traditional pharmacopeia of the ancient world, describing the uses of many garrigue plants such as euphorbia, hellebore, marrubium, germander, globularia and mandragora. It was in the Middle Ages that these plants were first taken from their hillsides to be planted in monastery gardens, and Charlemagne decreed in the *Capitulare de villis* a list of the plants to be grown in all gardens in the royal estates – a list that included sage, rosemary and rue. In the Arab gardens of Andalucia evergreen plants such as myrtle, which grows wild in the mountains above Granada, were often used to create clipped hedges, accentuating with their dark foliage the architectural lines of the interior courtyards; the Court of the Myrtles in the Alhambra is one example. Hundreds of years later, at the end of the 20th century, the landscape designer Nicole de Vésian focused on the different textures of garrigue plants to create sophisticated scenes where the plants, carefully clipped into topiary shapes, form successive planes that catch the light, as in her garden La Louve, now world-famous.

Top: A natural garden near Cape Saint Vincent in southern Portugal: thymes (*Thymus camphoratus*) grow mingled with cistuses (*Cistus ladanifer* var. *sulcatus*), germanders (*Teucrium vincentinum*) and cushions of spiny broom (*Stauracanthus genistoides*).

Bottom left: Common sage has been known since antiquity for its aromatic and medicinal properties. Here it is growing in its natural habitat, among stones on the Biokovo Mountains in Croatia.

Bottom right: Many wild garrigue plants are widely used in gardens. Here *Rosmarinus officinalis* and *Lavandula dentata* are seen in the garden of Rosie and Rob Peddle in Portugal.

On the Cycladic island of Antiparos, Thomas Doxiadis designs scenes where the transition from the cultivated garden to the wild landscape is often imperceptible.

Today, many landscape architects are inspired by the garrigue to create gardens that are not only beautiful but also require little water or maintenance. On the Greek island of Antiparos, for example, the landscape architect Thomas Doxiadis has designed a set of gardens that blend completely into the surrounding landscape, which is characteristic of the Cyclades. These gardens consist of garrigue plants with a compact and rounded habit that continues the natural flow of the surrounding vegetation so that the transition between the cultivated garden and the wild landscape sometimes becomes imperceptible.

Opening the boundaries of the Mediterranean garden so that garrigue plants are welcomed into it allows us access to a whole new world. Many wild plants of the Mediterranean are completely unknown to the majority of gardeners, who walk past this outstanding potential without realizing it. The flora of the regions surrounding the Mediterranean, to a great extent consisting of garrigue plants, is one of the richest in the world – it encompasses almost 25,000 species, in other words about a tenth of the flora of the whole world. Among these species are some real treasures for the gardener: the Cretan scabious (*Lomelosia minoana*), with monumental masses of silky leaves that spread over the south-facing slopes of Mount Dikti, making a magnificent groundcover full of thousands of mauve flowers in June; *Convolvulus oleifolius*, with delicate pink flowers, which forms a cascade of silvery balls below the temple of Poseidon on Cape Sounion, the very tip of Attica; *Globularia meridionalis*, with carpeting vegetation spiked with blue flowers that marries so well with the form of the rocks all along the Croatian coast; and Mount Etna broom (*Genista aetnensis*), with tree-like silhouettes, bending under the weight of their scented yellow flowers, tucked into the foot of the cliffs in the limestone gorges of eastern Sardinia. All of these plants seem to be waiting for gardeners to tame them and to benefit from their many qualities – flowering that is sometimes spectacular, a diversity of foliage colour, often amazing scents and a remarkable ability to withstand tough conditions, prolonged drought, strong winds and poor, stony soil.

• Different types of garrigue garden

The diversity of garrigue plants enables one to envisage gardens of very varied styles, far from the stereotyped image of the garrigue as a monotonous, dusty and austere landscape. In one area of our garden we have drawn inspiration from the natural landscapes of Cretan hillsides: the scene, in which shades of grey dominate, consists of a succession of cushion- or ball-shaped plants which allow a clear view of the Étang de Thau lagoon beyond. By contrast, near the house we have made the most of the ebullience of green-leaved plants – lentisks, myrtles, phillyreas, bupleurums and arbutuses – to delineate a space that is intimate and scented.

In one of his recent projects, the designer Jean-Jacques Derboux created a walled garden surrounding a long, minimalist-style pool, giving rhythm to this private space by the vertical lines of a few giant fennels, the shadow of their inflorescences against the walls evoking the pared-down graphics of a Japanese print. In Greece, the land-scape designers Jennifer Gay and Piers Goldson integrate into their many designs the plant palette of the nearby phrygana – *Ballota*, *Phlomis*, *Sarcopoterium*, *Thymbra capitata* (syn. *Coridothymus capitatus*), *Salvia pomifera* and *Salvia fruticosa* – to create subtle scenes of foliage with varying shades of green, grey and silver that remain attractive throughout the year. On the Côte d'Azur, the landscape designers Helen and James Basson are blazing new trails as they use their observation of the striking palette of colours in the garrigue to come up with unexpected scenes, lit up by the golden tones of wild grasses such as Coolatai grass (*Hyparrhenia hirta*), the bright orange of tree euphorbias (*Euphorbia dendroides*) and the deep rusty red of the dry inflorescences of *Cistus monspeliensis*.

• A new perspective on gardens

The garrigue garden invites us to discard our traditional aesthetic ideals. There are no lawns or old roses here, no beds of annuals smothered in flowers in mid-summer. This is instead a strongly seasonal garden with an appearance that changes profoundly between its flowering and dormant periods. Thanks to the structural character of evergreen foliage, the transitory flowerings seem almost an accessory. The garrigue garden is not subject to criteria that require a lush and flowery setting with the static perfection of a painting. It opens up a new perspective,

Top left: Many of the wild plants of the Mediterranean are little known to gardeners, who often walk past them without realizing their exceptional potential. Here the giant fennel of Catalonia (*Ferula communis* subsp. *catalaunica*) grows in the coastal garrigues of southern Portugal.

Top right: *Phlomis lychnitis* and *Convolvulus althaeoides* cover the ground between dwarf palms (*Chamaerops humilis*) on the coast south of Valencia in Spain.

Bottom left: *Lomelosia minoana*, a shrubby scabious that grows in the mountains of south-east Crete, flowers spectacularly in early summer.

different but just as interesting, which perhaps has more similarities with sculpture than painting: it is a space where volumes are emphasized by the continually changing play of light and shade, where the contrasts between foliage colours are heightened by alternating empty and filled spaces, open and closed areas, successive layers of tall plantings and of carpeting, all set in counterpoint with their mineral surroundings (stone, gravel).

Over and above its visual beauty, the garrigue garden offers us a complete multi-sensory experience: the rustling of wind in the leaves, the amazing silky texture of silver-leaved plants, the omnipresence of fragrances heavy on the warm air of summer evenings. Indeed, the garrigue garden is a place of discovery where the riches are inexhaustible: where plants are chosen as much for their ecological as their aesthetic qualities; where hoverflies dancing above fennel flowers or hungry caterpillars of the swallowtail butterfly on the pungent stems of rue are as precious as the plants on which they feed; and where the gardener may observe at close quarters the relationships between plants and these other small garden dwellers.

• *Horticultural considerations*

Making a garrigue garden is a thrilling experience, but for novices it can require considerable effort at first. This effort is primarily psychological, since all the instinctive horticultural principles that guide most gardeners throughout the world need to be swept away. While garrigue plants have exceptional survival strategies, they also have very particular demands that are quite different

from those of most cultivated plants. If they are to do well in the garden they require conditions that resemble their conditions in the wild as closely as possible.

Although this principle is simple, it may seem hard to apply since it involves going against a universal horticultural practice; instead of preparing the ground carefully before planting by improving it with organic matter to enrich it and retain moisture in the soil, we need to drain it by creating raised beds for our plants and to impoverish the soil as much as possible by, if need be, adding stones, pebbles or sand. Instead of encouraging growth by watering copiously in summer, we need to accept that garrigue plants should never be watered again once they have become established; we must let them develop according to their natural seasonal rhythm, which often includes a marked period of dormancy in summer to render them less vulnerable to disease or insect pests. Finally, instead

they weed their beds. Well-drained, poor and stony soil is ill-suited to the weeds that readily invade richer cultivated land and instead provides conditions that spectacularly favour the self-seeding of wild plants native to stony hillsides. In a garrigue garden, preparing the soil in this way has a remarkable twofold positive effect. First, by recreating conditions similar to those of the garrigue, we transform the garden into a dynamic landscape in which the planting evolves from year to year thanks to the unbounded generosity of euphorbias, cistuses, sages, phlomises, ballotas and coronillas, which self-seed freely through the beds. Second, the gardener takes on a new role: instead of battling against nature to maintain a fixed scene free of weeds, he or she learns to gently encourage the natural evolution of the garden. Once free of the eternally repetitive task of weeding, it's possible either to slow down or speed up the evolution of the landscape simply by deciding whether or not to keep the new plants that spring up every year after the first autumn rains.

In spite of its many advantages, the garrigue garden will never become the sole, or even the predominant, garden model because it works only when specific soil and growing conditions combine. If your garden is situated on rich, heavy and damp soil, it would be better to forgo the garrigue to avoid disappointment and turn instead to other options. A garrigue garden is just one of the possible ways of making a Mediterranean garden, completing the panoply of options available to the gardener. A garden can be designed to include several different but complementary zones adapted to different uses, with differing requirements for water and soil preparation and offering various emotional and sensual responses. The traditional vegetable garden, a guzzler of water and organic matter and thus very unlike the garrigue garden, has always had an important place in gardens in the Mediterranean, and between these two extremes there are numerous possible intermediate stages – for example, if you want abundant flowering in order to create a decorative summer scene in a strategic part of the garden, you could choose to limit

of giving the soil a mulch such as bark or shredded garden waste that will decompose in a few years to create a rich humus on the surface, we should apply an inorganic mulch such as gravel that will form a poor and perfectly drained surface better suited to dry-climate plants.

It isn't easy to discard basic horticultural rules, but the result is encouraging since it radically changes the maintenance requirements of the garden. An inorganic mulch to a great extent inhibits the germination of all those herbaceous annuals that gardeners spend a lot of time combating as

the irrigation required to satisfy the needs of thirsty flowering plants to a well-defined zone, like a mini-oasis. Nevertheless, in all regions with a dry summer climate the garrigue garden has incontrovertible advantages. Although little explored until now, it embodies a new approach which profoundly changes our relationship with nature. Quite apart from the incredible diversity of the plants it offers and the new aesthetic perspectives it opens up, it is also adapted to the challenges of the future: a garrigue garden enables us to reduce maintenance and minimize our use of both water and pesticides.

THE MEDITERRANEAN GARRIGUE:

A LANDSCAPE SHAPED BY HUMANS

Page 18: Sages, euphorbias, germanders, brooms, phlomises and ballotas on old abandoned terraces on Amorgos in the Cyclades.

Left: Goats were introduced into the Cyclades more than seven thousand years ago and now outnumber humans in maintaining the landscape.

Right: Goats sculpt the interlinked masses of kermes oaks into regular shapes like topiaries.

A dense network of *monopatia*, paths carefully paved to make it easier for mules to negotiate them, connects the villages and tiny blue-domed churches scattered along the length of the island of Amorgos, in the middle of the Aegean Sea. The paths are bordered on both sides by stone walls the height of a man, offering protection against the violence of the meltemi, the blustering north wind that blows during the summer months. On either side of the *monopatia* can be seen various aspects of a remarkable landscape that has been shaped by human activity over the course of several millennia. Countless dry-stone walls, testament to extraordinary stone-working skills, support the terraces that cling to the hillsides. On these narrow strips of land, rows of vines alternate with olive trees or minute fields of cereal crops, their boundaries marked by twisted almond trees or old figs rooted in the stone walls. Beyond the cultivated areas, the rocky landscape echoes with the constant tinkling of goat bells. Browsing ceaselessly on the toughest of leaves, the goats have sculpted the tangled mass of kermes oaks and wild olives into regular shapes, almost like topiary. Goats were introduced into the islands of the Cyclades more than 7000 years ago; on the small island of Amorgos, which has a human population of about 1900, more than 19,000 sheep and goats graze on the hillsides, transforming the steep slopes into a garrigue landscape that resembles a carefully pruned garden.

The *monopatia*, once trodden by the continual coming and going of mules, are now mainly used by hikers who come to explore the island during the springtime flowering of the phrygana, the local form of garrigue (see the box on p. 23). No one now toils to cultivate land situated far from the villages; profiting from this abandonment, the low-growing vegetation of the phrygana spreads over entire hillsides, with the memory of past agricultural activity still evoked by the regular traces of the dry-stone walls that follow the contours of the hills and give rhythm to the landscape. On the mountain that rises above the little port of Aegiali, the windmills which once stood proudly on a ridge exposed to the meltemi have fallen into ruin, invaded by clematis and surrounded by the inexorable advance of cushions of spiny burnet. Tree euphorbias, the foliage of which takes on magnificent yellow, orange, red or mauve hues in May, are among the first plants to self-seed on land that was formerly cultivated. They are followed by silver-leaved ballotas (*Ballota acetabulosa*), white-flowered oregano (*Origanum onites*), pink savory (*Satureja thymbra*), mauve germander (*Teucrium divaricatum*), strongly scented shrubby sage (*Salvia fruticosa*), bright yellow phlomis and cistuses with their crumpled petals; all these plants form the flowing structure of a landscape of extreme beauty, like a vast natural garden that has freely colonized the terraces descending towards the sea.

On a slope that fire has recently burnt, a carpet of bright green grass appears among the ashes when the first rains begin, rapidly spangled with the flowers of anemones, narcissi and crocuses which make the most of the newly created space to flourish among the charred rootstocks. By opening up the landscape and suddenly altering the structure of its vegetation, fire has left a profound mark on the landscape history of the island of Amorgos. Until the

19th century, a forest of deciduous oaks, the majestic *Quercus ithaburensis* subsp. *macrolepis*, had been exploited since antiquity for the edible acorns, whose cups, rich in tannin, were used in the tanning of leather. All that remains of this forest today are a few rare relics, standing isolated on Mount Krikellos amid sparse vegetation on stony soil, almost all the trees having been destroyed in 1835 in a terrible fire reported in the records to have burned for several weeks.

In the north of the Cycladic archipelago the island of Kea, lying close to Attica, is home by contrast to a large forest of oaks with edible acorns, the same species that has almost disappeared from the island of Amorgos. Renewed interest in the tannin contained in its acorn cups has

appeared recently, following the emergence in northern Europe of a demand for leather prepared without chemical products. Reviving a long tradition, nearly 35 tonnes of acorn cups were harvested in 2014, the harvest ending in a big fiesta at the port of Korissia, where the cargoes of acorn cups were traditionally loaded on to ships for export to the European continent. The new economic prospects opening up make it possible to envisage a return to ancient agroforestry practices on the island of Kea through the exploitation of a productive forest in which the ground under the trees, in part cultivated, is kept clear by the occasional passage of sheep. In an island which has preserved fine forests and where agricultural activity has been maintained, the garrigue is relegated to the poorest land, for example the slopes above the partially buried

In the Cyclades, tree euphorbias light up the landscape in May, self-seeding on what was once cultivated land of which the only traces remaining today are the dry-stone walls supporting the terraces.

Top: The Greek oak (*Quercus ithaburensis* subsp. *macrolepis*) was long exploited on the island of Amorgos for its enormous acorn cups rich in tannin. These oaks have progressively disappeared from the island as a result of recurrent fires.

Bottom: Agroforestry in the Cyclades where a mixed forest is dominated by oaks and the undergrowth is kept in check by the occasional passing sheep.

stage composed of young maples, Judas trees (*Cercis siliquastrum*) and turpentine trees (*Pistacia terebinthus*). This landscape, the structure of which resembles a garden reaching the first phase of maturity, becomes particularly beautiful in autumn when the new shoots of *Euphorbia dendroides* form a light green screen against which the splashes of yellow or orange of the Judas trees, turpentine trees and maples stand out.

theatre of the ancient city of Karthea. Here the recent evolution of the phrygana can be seen in two layers of vegetation, made possible by the progressive decrease in grazing. The rocky slopes, covered in tree euphorbias and sages, are punctuated by the development of a pre-forest

Garrigue, maquis or matorral?

The word 'garrigue' first appears in French in 1544, deriving from the Provençal name for the kermes oak, *garric*, itself possibly deriving from the pre-Roman word *carra*, stone. The garrigue is a landscape characteristic of the lands surrounding the Mediterranean Basin, dominated by shrubs and lower sub-shrubs, growing in arid conditions where the stony soil, usually limestone, generally remains partially visible between the plants.

The term 'maquis' is often used to denote vegetation equivalent to the garrigue growing on siliceous soil, as in the massifs of Les Maures and L'Estérel in southern France or in Corsica; the word derives from *macchia*, the Corsican word for the dense patches of vegetation that cover the hills. However, 'maquis' is also sometimes used in a different sense, to describe a type of vegetation on limestone soil that is taller than that of the garrigue, consisting of evergreen shrubs interspersed with holm oaks to form a dense cover. The Italian word *macchia* is also used for this tall cover of evergreen shrubs, whether they are on limestone soil or not, in contrast to the word *gariga*, which is applied to a low-growing type of vegetation consisting of plants such as rosemaries, euphorbias, cistuses and lavenders.

In other languages there are also distinct words for the different stages in the evolution of the landscape around the Mediterranean. In Greece, for instance, the word *phrygana* denotes the low-growing vegetation of landscapes that have had a long history of fires and grazing, with plants such as *Euphorbia acanthothamnos*, *Thymbra capitata*, *Ballota* and *Phlomis*, while the word *xerovouni*, sometimes used by ecologists, denotes tall vegetation consisting of myrtles, phillyreas, lentisks, wild olives and carobs. In Israel the word *choresh* is used for tall vegetation resembling that of the maquis, while the word *batha* denotes low-growing vegetation akin to that of the phrygana, where cushion-shaped plants and spiny plants capable of withstanding the extreme aridity dominate, such as *Sarcopoterium spinosum*. In Portugal, the *matagal* is a tall form of vegetation with partial tree cover, while lower-growing vegetation, called *mato*, includes cistuses, gorse and heather, for example. In the Algarve, in southern Portugal, three types of vegetation are distinguished according to soil type, altitude and proximity to the sea: *serra* on acid soil, *barrocal* on limestone soil and *litoral* on the area bordering the coast. In the Maghreb, the word *betha* denotes both tall maquis vegetation and the lower-growing vegetation of the garrigue, as distinct from steppe landscapes, called *sehb*.

The variety of words in different languages often conceals indeterminate borders, since the dynamic ecosystems of the garrigue evolve through numerous intermediate stages, independently of the soil type. In the 1960s the botanist Charles Sauvage suggested adopting the Spanish word *matorral* to designate more precisely the different stages of Mediterranean vegetation, regardless of the nature of the soil, with for example a low matorral if the plants are not more than 60cm (2ft) in height, a tall matorral if the plants are taller than 2m (6½ft), and an arboreal matorral if it includes trees. Nevertheless, in current usage the words 'garrigue' and 'maquis' remain the most widely used in both French and English, often in a very broad sense that covers different stages in the evolution of the vegetation and is sometimes unconnected with the nature of the soil.

Above from
left to right:
Phrygana near the
village of Lentas on
the south coast of
Crete: *Ballota
pseudodictamnus,
Sarcopoterium
spinosum* and
Thymbra capitata.

Maquis on the coast
west of Cap Corse:
*Myrtus communis,
Erica terminalis,
Rosmarinus officinalis,
Cistus salviifolius,
Calicotome villosa,
Phillyrea angustifolia*
and *Arbutus unedo.*

Garrigue on the
karstic screes of
Biokovo in Croatia:
*Euphorbia spinosa,
Iris pseudopallida,
Tanacetum
cinerariifolium* and
Pistacia terebinthus.

Spanish is without doubt the richest language for describing the diversity of garrigue landscapes. The different types of matorral are described by words deriving from the vernacular name of the principal species in the plant community: *romeral*, for example, is used for landscapes dominated by rosemary, often accompanied by lavender (*Lavandula dentata*) and heather (*Erica multiflora*), as on the flanks of the limestone mountain of Montgó south of Valencia; *tomillar* for low-growing vegetation with a mixture of thymes, helianthemums and santolinas, as on the crests of the Sierra de Cazorla in Andalucia; *jaral* for the sometimes impressive expanses of cistuses, as in the Albacete region where one sees hillsides entirely covered with the gum cistus (*Cistus ladanifer*); *sabinar* for the sombre landscapes colonized by *Juniperus phoenicea* subsp. *turbinata*, often sculpted by the wind or salt spray, as at Cap de Creus in northern Catalonia; *coccojar* for landscapes covered in kermes oaks, their impenetrable mass maintained by the cycle of recurrent fires; *lentiscar* for denser landscapes dominated by the shrubby vegetation of lentisks, viburnums, myrtles and arbutus; and finally *encinar* for a late stage of evolution, when the garrigues 'close up', giving way to holm oaks.

The great variety of words used to describe Mediterranean landscapes reflects the exceptional diversity of these plant communities, which evolve from one stage to another depending on complex interactions between the pressures exerted by human activity, the diversity of climatic conditions and the local particularities of soil and lie of the land. In this book I shall use the word 'garrigue' in as broad a sense as possible to denote the different types of vegetation that make up the remarkable mosaic of open Mediterranean landscapes, which the long history of human influences on nature has shaped to the point where, all around the Mediterranean Basin, they can be seen as the multiple facets of a vast natural garden.

• Dynamic landscapes

Over the centuries, Mediterranean garrigue landscapes have gone through phases of expansion or shrinkage according to the degree of pressure exerted on them by grazing, fire, tree-felling, land clearance and agriculture. The Western Mediterranean Basin, for example, shows two opposing tendencies. On its southern side, over-grazing constitutes a powerful driver behind the evolution of the landscape today, in some areas eliminating forests, garrigue and even the poorest steppes. In the Atlas Mountains the slopes above the village of Telouet, between Marrakech and Ouarzazate, are suffering from extraordinary erosion. The junipers, which elderly villagers say formed a forest that spread over all the mountains surrounding the village less than a hundred years ago, were cut down, opening the way to huge flocks of sheep which, day after day, searched out every last green shoot surviving between the stones. Prickly junipers (*Juniperus oxycedrus*), giving way higher up to magnificent incense junipers (*Juniperus thurifera*), were accompanied by a variety of sub-shrubs such as savory-leaved thyme (*Thymus saturejoides*), *Polygala balansae* with its purple and yellow flowers, and the tight cushions of *Bupleurum spinosum*. With the felling of the trees and the clearing of the vegetation, the roots that retained the soil were destroyed and the mountain is now creased into a multitude of fallen rocks and gullies that hurtle down to the Imarene wadi whose stony bed snakes through the country below, bordered by a narrow band of cultivated land. From time to time, witness to the vegetation of the past, a single old massive-trunked juniper stands alone on a mountain crest, protected through the centuries by its sacred virtue as a marabout tree, its dark windswept silhouette dominating the rare clumps of artemisia or euphorbia that still cling to the denuded slopes.

As you go back down towards Marrakech you pass through areas where a markedly weaker pressure from grazing allows a rich garrigue vegetation to thrive. At the edge of a forest protected as part of the royal estate, *Cistus laurifolius*, with its thick, glossy, leathery leaves, mingles with green bushes of stinking bean trefoil (*Anagyris foetida*) bearing bunches of acid-yellow flowers that appear hidden among the foliage in February, and silvery bushes of tree germander (*Teucrium fruticans*), of which the local variety has especially large flowers of a magnificent dark violet-blue.

By contrast, on the northern shores of the Western Mediterranean, in areas protected from urban sprawl, the abandoning of pastoral activities has led in a few decades to a return of woodland dynamics, with the garrigue closing up to form a dense environment into which light can no longer penetrate. Near the Étang de Thau, on the hillsides closest to our nursery, the ancient garrigue paths now wind through dense woods where holm oaks emerge from a tangle of viburnums, phillyrea, lentisks, honeysuckle and Mediterranean buckthorn. These dark environments, frequented by wild boar, are sometimes made completely impenetrable by the tangled thorny stems of sarsaparilla, which climbs into shrubs then hangs from the branches of the tallest oaks in long sheets, covered in late summer with small white flowers whose honey-like scent attracts pollinators.

In these environments there is less and less room for garrigue plants. The only open spaces are sometimes the edges of fire breaks where the regular passage of rotary slashers enables grassy expanses of brome to grow, lit up in early spring by the yellow, white, blue or purple flowers of the dwarf iris (*Iris lutescens*). The ruts along the centre of the tracks left by the wheels of hunters' vehicles are also often the site of amazing miniature gardens where one may see the delicate bell-shaped flowers of yellow flax, the intense blue of gromwells and the finely cut leaves of mountain rue or calaminth, whose fresh scent is released as soon as the huge tyres of all-terrain vehicles brush against them. In clearings where the very poor soil delays the establishment of shrubby species, the wild boar come at night to turn over the stones between *Sedum* species,

thymes, helianthemums, snapdragons, *Lavandula spica*
and *Phlomis lychnitis*, rooting through the soil in search
of the bulbs of the elegant wild tulips, jonquils, woodcock
orchids and asphodels (*Asphodelus macrocarpus*), all of
which plants need light and thus make the most of areas
not yet covered by vegetation.

In the South of France, the return of holm oak forests at
the expense of the garrigue is speeded up by the impressive
advance of the Aleppo pine, ready to take over indiscriminately the margins of housing developments, abandoned

fields, wasteland or former garrigue. In the heart of the
Department of Hérault, the cirque or blind valley of
Mourèze, famous for its ruiniforme lunar landscape, has
for a long time been the home of an especially beautiful
garrigue. Among the curiously shaped rock formations,
the very well-drained soil resulting from the erosion of the
dolomitic rock has favoured a great diversity of perennials
and sub-shrubs: *Aphyllanthes monspeliensis* and *Linum
narbonense*, helianthemums and *Teucrium aureum*, *Cistus
salviifolius* and helichrysums, *Euphorbia nicaeensis* and
Daphne gnidium, *Globularia alypum* and *Erica multiflora*.

Left: The garrigue reconquers abandoned terraces in Portugal.

Right: The abandoning of pastoral activities has changed the forest dynamic in just a few decades as advancing pines drive out cistuses.

Bottom: Pines and cypresses colonizing former coastal garrigues north of Dubrovnik, Croatia.

The Mourèze cirque, which is like a magnificent rock garden extending over tens of hectares, was for a long time shaped by two activities. Up until the Second World War the systematic cutting of wood by charcoal-burners removed the forest, pushing the oak copses back to the heights of Mount Liausson, which separates the Mourèze cirque from Lake Salagou. Completing the charcoal-burners' work, grazing by goats ensured that brushwood

Right: A clearing on the edge of a garrigue on the Larzac plateau in the process of becoming closed. *Thymus vulgaris* and *Helianthemum oelandicum*.

Opposite left: Now that goats have disappeared and charcoal burning no longer takes place, the Mourèze cirque is undergoing colonization by pines.

Opposite right: Progressive colonization by pines increases the mass of flammable vegetation with the result that fire is becoming one of the main forces acting on ancient garrigues on the northern side of the Mediterranean.

Opposite bottom: A young garrigue in southern Spain flowering three years after a fire: *Cistus monspeliensis, Phlomis lychnitis, Gladiolus italicus, Teucrium pseudochamaepitys, Helianthemum syriacum* and *Lavandula dentata*.

in the cirque was regularly cleared, favouring some rare species that require open land and perfectly drained soil, such as the miniature cushions of cranesbill (*Erodium foetidum*) or the graceful tufts of grass-leaved daisies (*Leucanthemum graminifolium*). The human activities of charcoal-burning and grazing created a unique environment here, home to a particularly rich flora and fauna, which has recently been listed as a heritage site. Today, however, with the disappearance of grazing goats and the abandoning of charcoal-burning, the Mourèze cirque is undergoing spectacular colonization by pines. As the dolomitic rock formations are cloaked by these trees, shade, root competition and the layer of pine needles that carpets the soil are little by little driving away the plants belonging to an open environment. The growth of the pines adds considerably to the biomass of inflammable vegetation; the probably inevitable outbreak of a fire may be the next force to act on the landscape, opening up the space to offer the plants of the garrigue a temporary expansion phase once more.

• *Landscapes in equilibrium*

For thousands of years, in a slow process accelerating sometimes in one direction and sometimes in another, garrigue landscapes have evolved between an open environment and a progressive closing of the vegetation which leads, when climate and soil conditions permit, to the re-establishment of forest cover. This evolutionary cycle of the garrigue, however, is not always a case of continual movement.

It is sometimes marked by periods of stability; depending on environmental conditions and the history of disturbances linked to human activity, it sometimes happens that the evolutionary trajectory becomes fixed and the ecosystem remains stable for a longer or shorter period before ultimately starting off again on its evolutionary course. The landscape enters into a phase of equilibrium, the plant dynamics adopting a cycle that is apparently static, like a spinning top whose vertical axis remains for a moment on a fixed point. The array of species adapted to a specific environment is replenished as the plants grow old, the vegetation renewing itself without progressing to a subsequent stage. These landscapes in equilibrium are of particular interest from the gardener's point of view, for they can serve as an inspiration for gardens that require little maintenance: when it is in a stable phase the garrigue offers the perfect model for a garden without a gardener.

In northern Mallorca, the slopes of the mountains dominating the bay of Cala Sant Vicenç, shaped by their history of a particular use of fire, today resemble a landscape in equilibrium, like a vast garden that maintains itself. The vigorous grass *Ampelodesmos mauritanicus*, common in Mallorca, has long ribbon-like leaves with finely serrated edges that protect the plant very effectively, preventing sheep from eating them. Unlike old leaves, however, the tender new shoots which regenerate after fire are much appreciated by sheep as they have not yet developed their serrations. Exploiting this ability of the plant to produce tender new leaves after having been burnt, for centuries shepherds set fire to the hillsides every year to create pastureland, thus progressively transforming the landscape. By eliminating species which cannot withstand frequent fires they favoured the growth of *Ampelodesmos* to the point of creating an almost monospecific landscape, ideal for intensive grazing because of the profusion of young shoots regularly renewed by the flames.

Once a simple fishing village, Cala Sant Vicenç has been progressively transformed into a cosmopolitan tourist centre and the landscape has evolved along with it. Today the hillsides are no longer burned by shepherds: cultural perceptions of fire have changed, and it is no longer seen as a tool for managing the landscape but rather as a danger to be combated. The few sheep that still roam the mountains these days pick their way through the great clumps of *Ampelodesmos*. They avoid its now unpalatable long leaves and turn instead to other plants, such as the bushes of wild olive which they browse on patiently until they have transformed them into a curious carpet which marries perfectly with the shape of the rocks. Without regular fires the garrigue has become enriched with numerous species ignored by sheep: pink-flowered germanders (*Teucrium capitatum* subsp. *majoricum*), the dark-leaved *Cneorum tricoccon*, the Balearic buckthorn (*Rhamnus ludovici-salvatoris*), the Balearic St John's wort (*Hypericum balearicum*), pink-flowered phlomis (*Phlomis italica*), joint pine (*Ephedra fragilis*) with its interweaving stems and Balearic boxwood (*Buxus balearica*), whose tough vegetation is often ground-hugging, levelled by the salt spray carried on the tempestuous north wind. Today this very beautiful garrigue is in a temporary state of equilibrium as a result of the different forces that affect plants, with species self-seeding at will into free spaces and without any notable evolution towards another stage. In time, the landscape will begin to evolve dynamically again: the Aleppo pines that are already present higher in

A landscape in equilibrium under the influence of grazing: sheep pick their way between thymes, heathers and lentisks on the Rodopou peninsula in Crete.

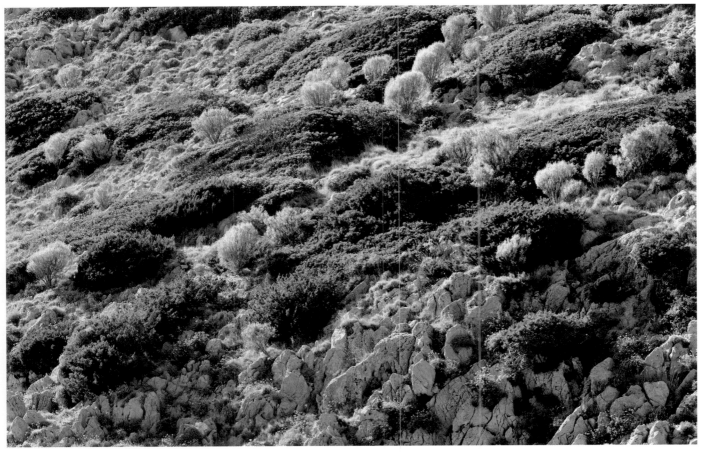

Top: A landscape in equilibrium on a rocky slope by the sea on the island of Capri: a carpet of lentisks flattened by wind and salt are growing with Phoenician junipers and the white balls of Jove's beard (*Anthyllis barba-jovis*).

Bottom: A mountain landscape in Cala Sant Vicenç, Mallorca: wild olives hug the ground in response to pressure from grazing. The sheep and goats reject the grass species that are no longer palatable now that the cycle of burning has been discontinued.

Balearic boxwood (*Buxus balearica*) flattened by salt spray.

the Serra de Tramuntana may once more trigger the fire cycle, or may instead open the way to a slow return to mixed forest dominated by oaks.

With its variety of climatic conditions and its long history of human influence on nature, Crete contains a large number of landscapes in equilibrium, fascinating to the gardener. The Gramvoussa peninsula, for example, which forms a narrow mountainous spur in the extreme north-west of the island, has had a history similar to that of Cala Sant Vicenç in Mallorca, although its flora is completely different. Having always been devoted to grazing sheep and goats, the Gramvoussa peninsula has no buildings on it, which has allowed fire to be used for centuries to open up the landscape on this long strip of isolated land. Repeated fires have entirely eliminated trees and shrubs.

Only a few rocky ravines, not easy to burn, still shelter dark flows of carob trees which down by the sea give way to dense masses of *Vitex agnus-castus* and oleanders. The rest of the landscape consists of an extraordinary collection of cushions and balls of vegetation, which from a distance look as if they are rocks. The dominant plant here is the Cretan thyme, *Thymbra capitata* (syn. *Coridothymus capitatus*), whose very regular ball shape is covered in pink flowers during the long, scorching summer months, offering abundant nectar to bees. During these months too the arid hillsides take on their summer dress – countless rows of multicoloured beehives.

At the tip of the peninsula is the famous beach of Balos, which attracts thousands of visitors every year since it is held to be one of the most beautiful in the Mediterranean

on account of its circular shape surrounding a natural lagoon. Every day in summer the bumpy track leading to Balos sees a long line of slowly moving cars following each other in a thick cloud of dust. Tourism has brought about a shift in economic activity and the pressure of tourism has limited the use of fire and reduced the intensity of grazing, with a secondary effect on the vegetation. The landscape of the peninsula has now entered an equilibrium phase, with a particularly beautiful phrygana in which the thyme is accompanied by numerous other species. Contrasting with the silver mass of ballotas (*Ballota pseudodictamnus*), the glossy leaves of lentisks extend in vast ground-covering carpets, flattened by salt spray in the same way as the Balearic boxwood on the northern coast of Mallorca. Between the thymes and ballotas and lentisks grow many other plants whose cushion shapes emerge among the stones: spiny burnet (*Sarcopoterium spinosum*), dark green buckthorns transformed by goats into perfect spheres (*Rhamnus lycioides*), broom covered in yellow flowers in spring (*Genista acanthoclada*), Jerusalem sage (*Phlomis fruticosa*) with leaves that are green, grey or golden according to the season, and lime-tolerant heather with a spectacular flowering in autumn (*Erica manipuliflora*). Here, as so often around the Mediterranean, it is the combination of aridity, the salt borne on the vigorous sea spray, the poor, stony soil and reduced grazing that maintains this plant community in a stable state, forming a landscape garden which the tourists who cross the pen-

insula admire as a wonderful natural landscape without ever suspecting that it has been entirely shaped by man.

Depending on the history of disturbances caused by human activity, the evolution of the vegetation has followed a variety of distinct trajectories which go to make up the extraordinary diversity of garrigue landscapes that we see all around the Mediterranean. Whether they are in phases of change or in periods of stability, these landscapes can serve as models for gardeners; the creation of a garden may be inspired by the natural dynamics of garrigue vegetation, its potential evolution prompting us to consider how our initial planning will develop over the longer term. The way we manage maintenance can be informed by the ability of garrigue landscapes to enter into equilibrium phases: the garden then requires minimal maintenance which, on a smaller scale, should replicate some of the external forces that influence the vegetation on the greater landscape. When we look at Mediterranean landscape with a gardener's eye, we may start to conceive of a garrigue garden as an ecosystem, the transformation of which the gardener can guide over the years and in doing so bring its dynamic evolution closer to the workings of the landscapes that inspired it.

Top: The Gramvoussa peninsula has always been used for grazing sheep and goats and therefore has no buildings on it, which has allowed the unrestricted use of fire to keep the landscape open on this long, isolated strip of land for centuries.

Bottom: Looking at natural landscapes of the Mediterranean nature with a gardener's eyes, one can visualize a garden where the constituent elements of an ecosystem pass through periods of evolution and periods of equilibrium.

Following pages: A landscape in equilibrium on the Gramvoussa peninsula, Crete. The combined action of drought, salt carried by spray during storms, the poverty of the stony soil and moderate grazing currently maintains this landscape in a stable phase.

'LANDSCAPE GARDENS' OF THE MEDITERRANEAN

As we walked along the coastal path in south-west Crete in search of Kedrodassos, known for its ancient twisted junipers growing among dunes, Clara and I suddenly stopped, struck by the beauty of the scene before us. In the morning light of a November day the silvery cushions of thyme and the green balls of spiny broom, the expanses of pink heather, the counterpoint of limestone rock showing between the plants, the vertical rhythm of the tall dried flowering heads of sea squill and the dark foliage of carobs which closed off the view and gave a sense of intimacy to the scene – all these architectural elements of the landscape seemed to form the framework of a magnificent garden. That moment when we paused, the moment when a gardener is moved by a landscape, was not an isolated experience. Indeed, in the course of our travels Clara and I have amassed in our memories a collection of such scenes where the garrigue could have been a garden. 'Landscape gardens' abound in Crete, in the mountains of the Peloponnese and in the Cycladic islands, which for us remain an inexhaustible source of inspiration. But you can also find them all round the Mediterranean Basin, in the red folds of ruffe (a red sedimentary rock with erosion patterns) on the banks of Lake Salagou, on our own door-step, as well as in the expanses of *Salvia tomentosa* spreading beneath the cedars on the crests of the Alawite mountains at the other end of the Mediterranean.

On the Larzac plateau in the south of the Massif Central in France, a simple stony clearing by the side of the road, used as a resting point by tourists on their way to the fortified village of La Couvertoirade, is colonized by a remarkable group of rock plants: carpeting thyme, *Teucrium aureum*,

Anthyllis montana, Astragalus monspessulanus, common helichrysum and *Aphyllanthes monspeliensis*; emerging from them are the delicate flowering stems of salvias and the swaying plumes of *Stipa pennata*. Behind the fuel reservoirs of a filling station on the E65 motorway in Croatia that bypasses Split and climbs towards Zadar, the expanse of white karst is softened by the rhythm of the perfect balls of euphorbias (*Euphorbia spinosa*) regularly scattered among the cushions of *Thymus longicaulis,* on to which in the evening the blue petals of *Linum perenne* fall. In the ruins of the ancient city of Phaselis in Turkey, the sculpted stones and fragments of marble columns strewn on the ground disappear beneath the exuberant vegetation of acanthuses, cyclamens and red-trunked arbutuses which form the structure of a shade garden beneath the canopy of the pines that have invaded these ruins by the sea. Between

Lorca and Guadix, in one of the driest regions of Spain, the hillsides are covered as far as the eye can see in the golden undulations of *Anthyllis cytisoides*, punctuated in spring by the flowering stems of blue-mauve viper's bugloss (*Echium*

vulgare) and greeny-yellow giant fennels (*Ferula communis*) in an extraordinary garden scene that extends for kilometres all round.

In the landscapes that we study around the Mediterranean it is not only the aesthetic aspect that interests us, attractive though it is, but also a more global understanding of the way the ecosystem works, for it is this that makes possible the garrigue's unexpected gift to us: the spontaneous creation of a garden without a gardener. The particular climate and soil conditions, the local history of a landscape's evolution, the different forces that affect an ecosystem to maintain it in a phase of temporary equilibrium, its dynamic potential and its probable future on a time scale from a few years to a few decades: these are all worth studying and may inspire our reflections on garrigue gardens that require little maintenance. To apply this twofold approach, aesthetic and functional, I have chosen ten symbolic garrigue landscapes from different regions around the Mediterranean, inviting you to travel to them with me in the following pages. In rainy or dry climates, on alkaline or acid soil, by the sea or near mountain passes, these landscapes illustrate the extraordinary diversity of plant communities that could inspire Mediterranean gardeners to recreate garrigue gardens, as varied as possible, in their own particular environments.

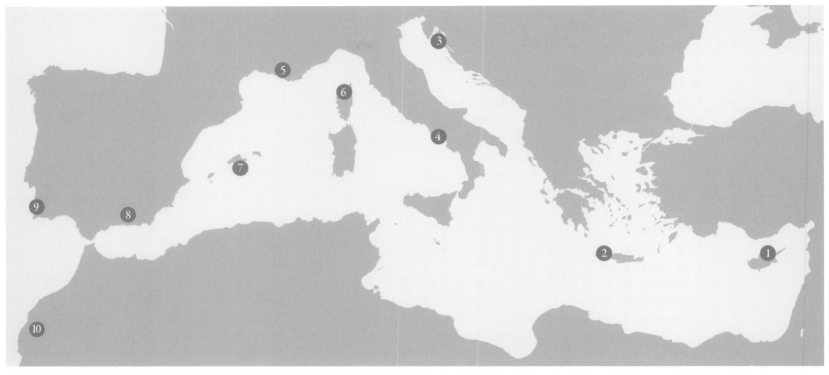

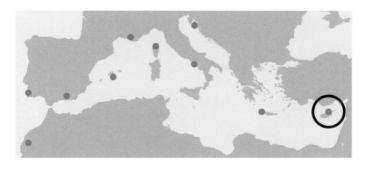

Cyprus: between mountains and peninsulas

Cut off by the sea for several million years, the island of Cyprus is home to a striking flora that includes many endemic species. In the Troodos Mountains in the centre of the island, branching bushes of the golden oak (*Quercus alnifolia*) grow beneath old tabular cedars (*Cedrus libani* var. *brevifolia*). In the west, the limestone hills of the Akamas peninsula are covered with a low-growing garrigue in which shrubby sage (*Salvia fruticosa*) and silver-leaved phlomis (*Phlomis cypria* var. *occidentalis*) mingle with a remarkable diversity of natural cistus hybrids. It is, for example, from the Akamas Peninsula that *Cistus × skanbergii* comes, a spontaneous cross between *Cistus monspeliensis* and *Cistus parviflorus* which has become one of the most sought-after cistuses for Mediterranean gardens because of its abundant pale pink flowers. At the other end of the island, the Karpas Peninsula is a narrow strip of land that extends for more than 70km (43 miles), where herds of wild asses frolic among the junipers along the coast which bear witness to the recent evolution of the landscape.

The sudden abandonment of a large part of traditional agro-pastoral activities following the partition of the island in 1974 brought about a rapid transformation of the garrigues in northern Cyprus. On the slopes of the Kyrenia Mountains along the coast opposite Turkey the dynamics of vegetation that has been left to itself for several decades, after centuries of fire and grazing, have allowed the return of mixed forest in which different layers of plants grow in proximity. The lowest layer forms a rich garrigue where cistuses, giant fennel (*Ferula communis*)

and Cretan germanders (*Teucrium creticum*) grow in the midst of the thistle-like, velvety-leaved *Ptilostemon chamaepeuce* and *Prasium majus* with its dark green leaves. Beneath the shrubs and among the stones, the fragrant flowers of the Cyprus cyclamen (*Cyclamen cyprium*), their narrow white petals blotched with purple, surround the vigorous ground-hugging rosettes of mandrakes (*Mandragora officinarum*). In the intermediate layer, the magnificent red, ochre or mauveish trunks of the Cyprus arbutus (*Arbutus andrachne*) are sometimes accompanied by storax (*Styrax officinalis*) or Syrian maples (*Acer obtusifolium*), the evergreen foliage often lightly draped with wild clematis (*Clematis cirrhosa*). The upper layer is made up of the twisted shapes of cypresses and Calabria pines (*Pinus brutia*), which colonize the poorest areas of the limestone slopes and continue up to the peaks, where a few ruined castles dating from the time of the Crusades still stand.

On the rocks around their walls a particularly beautiful collection of species is to be found, which transform the ruins into rock gardens. *Sideritis cypria* spreads among the blocks of stone in large, downy white cushions that in winter blend in with the colour of the rocks, while by contrast in spring the plant stands out for its spectacular inflorescences with red stems and lime-green bracts. Shrubby silene (*Silene fruticosa*), the sticky leaves of which grow directly from cracks in the walls, is covered in bright pink star-shaped flowers. The spiral inflorescences of *Onosma giganteum* open to reveal golden-yellow flowers bristling with silky hairs. The multicoloured gromwell (*Lithodora hispidula* subsp. *versicolor*) self-seeds everywhere among the rubble, its rough leaves disappearing beneath a mass of tubular flowers in pink, mauve and white.

Sometimes, near the tops of the mountains, like a treasure offered by the garrigue, a discreet bush of the Cyprus oregano (*Origanum majorana* var. *tenuifolium*) grows among the gromwells, cistuses and sages, recognizable by its velvety grey leaves that are deliciously aromatic, giving off what is without doubt one of the most subtle and pervasive scents of all Mediterranean plants.

1. The Cyprus arbutus (*Arbutus andrachne*), with its magnificent trunk that is orange-red or mauveish depending on the time of year, is anchored in the limestone rocks of the Kyrenian mountain chain in the north of the island.

2. Herds of donkeys that have become wild in the Karpas peninsula are a reflection on how some agricultural and pastoral activities have been abandoned.

3. Cypresses and Calabria pines (*Pinus brutia*) in their full dynamic reconquest of garrigues where sheep are no longer grazed.

4. The limestone peaks of the Kyrenian mountain chain are full of endemic species such as *Sideritis cypria,* whose white foliage blends in with the colour of the rock.

5. The rare shrubby silene (*Silene fruticosa*) finds a home in the walls of castles dating from the time of the Crusades, which stand on the peaks of the Kyrenian mountains.

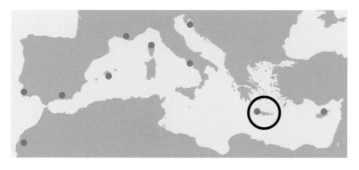

Crete: the White Mountains and gorges that descend to the Libyan Sea

The landscapes of Crete are full of sound, thanks to the continual tinkling of sheep and goat bells that echo in the stony valleys. The total number of sheep and goats on the island today amounts to more than a million animals which patiently sculpt the garrigue as they browse on the tiniest tender shoots, transforming the kermes oaks and wild olives into thousands of strangely shaped topiaries worthy of a fairy tale. Far from impoverishing plant diversity, the long history of grazing in Crete has shaped a unique cultural landscape, remarkably diverse, which can serve as a model for a 'landscape garden' rich in numerous species. At the foot of the White Mountains, in the dry coastal strip bordering the Libyan Sea, phrygana species are perfectly adapted to withstand the pressure of grazing. The tiny leaves of heather (*Erica manipuliflora*) and of anthyllis (*Anthyllis hermanniae*), delicacies for goats, are rendered inaccessible to them by the hard tips of the tightly packed branches that protect the heart of the plants. Some species such as asphodel (*Asphodelus ramosus*) and euphorbia (*Euphorbia characias* subsp. *wulfenii*) contain toxic compounds in their sap which protect them from herbivores. The many strongly aromatic plants, such as *Salvia pomifera*, summer-flowering thyme (*Thymbra capitata*, syn. *Coridothymus capitatus*) and pink savory (*Satureja thymbra*), are only browsed as a last resort when the goats can find nothing else to eat. Other cushion-shaped plants, such as spiny mullein (*Verbascum spinosum*), spiny burnet (*Sarcopoterium spinosum*), centaurea (*Centaurea spinosa*), downy ballota (*Ballota acetabulosa*) and woolly

phlomis (*Phlomis lanata*), shelter beneath the protection of spines or dense hairs which colour the phrygana in shades of gold and silver while at the same time making the plants unpalatable to sheep and goats.

Species that are vulnerable to grazing have chosen to leave the phrygana and find refuge on inaccessible rockfaces in the many limestone gorges that cut through the southern slopes of the White Mountains. Densely populated with striking species, often extremely beautiful, the rocks resemble vertical botanic gardens. In the Aradena Gorge, reached by crossing a vertiginous bridge, bright pink *Ebenus cretica* grows in the fissures among straw-coloured centaureas (*Centaurea argentea*) and the amazing *Petromarula pinnata*, with extravagant infloresences that bear thousands of bell-shaped pale blue flowers. In the Kotsifos Gorge, *Iris unguicularis* subsp. *cretensis* threads through the clumps of the glaucous *Linum arboreum* and the heavy hanging balls of leathery-leaved *Staehelina petiolata*, while at the bottom of the gorge, tucked between huge blocks of stone, are little silky clumps of dittany of Crete (*Origanum dictamnus*), their scent piquant and spicy.

Higher up, along the path leading down from the Omalos plateau, crossed by flocks of sheep during transhumance, the white cushions of *Sideritis syriaca*, famed for its medicinal properties, grow among loose stones along with *Helichrysum italicum* subsp. *microphyllum* and mountain oregano (*Origanum microphyllum*), whose miniature leaves are masked by a mist of mauve flowers in summer. On the mountain crests the spiny cushions of the magnificently architectural *Euphorbia acanthothamnos* self-seed at the foot of monumental cypress trees whose thick trunks, twisted by wind, snow and cold, have undergone centuries of natural pruning. On the cliffsides, scabiouses with white-margined leaves (*Lomelosia albocincta*) bear large mauve flowers followed at the end of summer by dry seedheads resembling decorative pompons; their paper-thin sections allow the seeds to exploit the slightest air currents rising from the hot rocks to carry them into ever higher crannies, safe from the rasping tongues of goats.

1. In Crete more than a million sheep and goats sculpt the garrigue by browsing on even the tiniest green shoots.

2. Below the imposing mass of Mount Gingilos, which reaches a height of more than 2000m (6562ft), a spiny cushion of euphorbia (*Euphorbia acanthothamnos*) grows beneath a cypress whose twisted trunk is the result of centuries of coppicing.

3. The bright pink flowers of *Ebenus cretica* decorate the sides of numerous gorges in the mountains of Crete.

4. The strongly scented essential oils of *Salvia pomifera* protect the plant from sheep and goats.

5. Mountain tea or malotira (*Sideritis syriaca*) is used in Crete for its numerous medicinal properties.

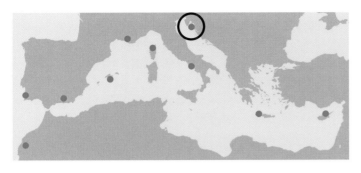

Dalmatia: karst landscapes on the Adriatic coast

The Biokovo Mountain seems to plunge vertically into the sea. Rising to more than 1700m (5577ft) above the Adriatic, this massif with its jagged outline presents a succession of white cliffs and rocky plateaux forming a landscape that looks like a semi-desert. The white karst, characteristic of Croatian landscapes, is riddled with countless holes, fissures and crevices. The word karst, derived from the name of the region of high limestone plateaux in Slovenia and Croatia, denotes a landscape with a particular structure resulting from changes in the rock caused by the chemical action of water. Rainwater becomes acid when it is charged with carbon dioxide, first in the atmosphere and then on contact with the humus in the soil, and slowly eats into the underlying limestone. This process then forms more or less parallel dissolving channels on the surface of the karst, sometimes jagged and fretted with sharp protuberances, while beneath the surface it creates a complex network of cavities, caves, abysses and underground rivers. The outline of the karst is often invisible, hidden beneath the earth and vegetation.

The spectacular karstic landscapes that we see today on the mountainous coast of Croatia and on the numerous islands scattered off its shores have been stripped bare by a long process of erosion, caused by the double action of over-grazing and the violent rains to which the region is prone. The chain of the Dinaric Alps along the coast effectively blocks the influence of the sea, giving rise to rainstorms of unusual intensity on the Dalmatia coast. Today these striking rocky landscapes are slowly being transformed into natural gardens where one can observe how well Mediterranean vegetation adapts to a damp climate in stony soil with perfect drainage. The decline in agriculture since the conflicts of the 1990s has led to a rapid reduction of grazing, allowing plants to start taking over the immense expanses of rocky ground. On the island of Pag, in spite of the blustering prevailing winds, the bora, the maestral and the jugo, which sweep salt on to the denuded coasts, the stony screes are colonized by the magnificent yellow inflorescences of asphodelines (*Asphodeline lutea*). On the island of Krk, the vast white screes have been invaded by aromatic clumps of sage (*Salvia officinalis*) and thyme (*Thymus longicaulis*) with bright pink flowers and long stems that weave between the blocks of stone. On the slopes of Biokovo, once constantly traversed by sheep and goats, the white daisy flowers of pyrethrum (*Tanacetum cinerariifolium*) grow in crannies between cushions of euphorbia (*Euphorbia spinosa*), dense carpets of globularia (*Globularia meridionalis*), the downy rosettes of mullein (*Verbascum thapsus*), the dark stems of savory (*Satureja montana*), robust clumps of the Dalmatic iris (*Iris pseudopallida*) and the tall inflorescences of *Campanula pyramidalis*.

To the north of Zadar, the rocky slopes descending to the sea are home to a young forest, curious-looking and very beautiful, like a lunar forest landscape. The widely spaced trees that have managed to germinate in the deepest fissures grow among enormous blocks of stone, with no intermediate vegetation since no other plants have succeeded in establishing themselves in the compact rocks under the trees. In November, the red, yellow and orange leaves of maples (*Acer monspessulanum*) and light brown of downy oaks (*Quercus pubescens*) dominate the landscape in contrast with the dazzling white of the limestone. In April, the dark green of phillyreas (*Phillyrea latifolia*), the translucent red of the tender new leaves of turpentine trees (*Pistacia terebinthus*), the luminous yellow of the young shoots of hackberries (*Celtis australis*), Christ's thorn (*Paliurus spina-christi*) and hop-hornbeams (*Ostrya carpinifolia*) mingle with the creamy white of ash trees (*Fraxinus ornus*), weighed down by their panicles of flowers whose honey scent floats on the air along the long stretches of rocky coast.

1. Forest in a karst landscape north of Zadar: *Phillyrea latifolia, Fraxinus ornus, Acer monspessulanum* and *Pistacia terebinthus.*

2. *Asphodeline lutea* growing on a limestone scree on the island of Pag.

3. *Stipa calamagrostis* flowers in summer in the Biokovo Mountains.

4. Garrigue with sage (*Salvia officinalis*) and a turpentine tree (*Pistacia terebinthus*).

5. *Globularia meridionalis* forms a low carpeting groundcover that spreads among the stones.

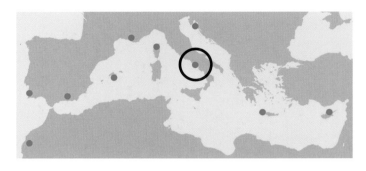

The Amalfi peninsula: garrigue and agricultural terracing

The Amalfi peninsula is a long, mountainous stretch of land jutting out into the Tyrrhenian Sea and closing off the Bay of Naples. The Amalfi coast, classified by UNESCO as a World Heritage Cultural Landscape, is furrowed by a succession of steep valleys, the landscape of which has been meticulously worked by humans since the Middle Ages. Vines, olives and lemon trees are cultivated on thousands of terraces designed to retain the soil on the steep slopes above the sea. Vesuvius lies nearby: during the eruption of 79CE, which buried the towns of Herculanum and Pompeii under a thick layer of ash, the limestone massif of the Amalfi peninsula was also covered by several metres of volcanic debris. The soil, composed today of volcanic soil modified by erosion and mixed with the underlying limestone, is fertile, rendering sophisticated cultivation techniques possible. To make the most of the space on the narrow terraces, vines are grown on tall pergolas and lemon trees are protected by large shade frames carefully constructed from chestnut poles, on which the *contadini volanti*, or 'flying farmers', balance to prune, stake and harvest their crops.

A network of narrow paths, often in the form of steps at their steepest points, links the cultivated terraces to the villages beside the sea, where the harvested lemons were traditionally taken aboard ships to prevent scurvy. The local variety of lemon, the Amalfi *sfusato*, is famous for its exceptional vitamin C content. Above the topmost cultivated terraces the paths go on climbing to the forests of the high plateaux, situated at an altitude of more than 1000m (3280ft), where the chestnut trees that thrive on the thick layer of volcanic soil are regularly cut and coppiced to provide the straight poles needed by farmers.

Crossing a rich mosaic of landscapes where cultivated land alternates with garrigue, the paths along the Amalfi coast allow one to discover a remarkably diverse flora.

About 900 plant species grow on this narrow peninsula. The dry-stone walls are home to snapdragons with bi-coloured flowers (*Antirrhinum siculum*), grey-leaved stocks (*Matthiola sinuata*) and heavy cascades of capers bearing white flowers with large bunches of mauve stamens all summer long. The steps that descend from terrace to terrace are like miniature gardens, with the delicate white flowers of crocuses (*Crocus imperati*) preceding scented narcissi (*Narcissus tazetta*), the white umbels of ornithogalums (*Ornithogalum umbellatum*) and grape hyacinths (*Muscari botryoides*). Between the fields the rocky outcrops are colonized by pink cistuses (*Cistus creticus*), santolinas with finely cut leaves (*Santolina neapolitana*), silky balls of silver convolvulus (*Convolvulus cneorum*), tight cushions of a scabious with small crenellated leaves (*Lomelosia crenata*, syn. *Scabiosa crenata*) and the magnificent rosemary-leaved gromwells with bright blue flowers decorating the dark foliage (*Lithodora rosmarinifolia*). On inaccessible ledges in the cliffs, holm oaks make little patches of woodland as they cling to the rock. At the bottom of cooler valleys, in the shade of myrtles and honeysuckle, the large yellow-flowered sage (*Salvia glutinosa*) grows alongside irises (*Iris foetidissima*) whose fruits open in autumn to reveal fleshy seeds, each in a crimson seedcoat.

On the poorest coastal land, once used for grazing, as on the slopes of Punta Campanella which faces the island of Capri, recurrent fires have led to the creation of vast golden steppes where grasses dominate (*Hyparrhenia hirta*), out of which emerge the tall flowering stems of giant fennel (*Ferula communis*) and the massive, perfectly spherical shapes of tree euphorbias (*Euphorbia dendroides*). The cliffs by the sea are home to plants that are clad in dense hairs in order to better withstand the salt spray: twisted bushes of Jove's beard (*Anthyllis barba-jovis*), compact cushions of helichrysum (*Helichrysum litoreum*) and generous clumps of *Centaurea cineraria*, whose bright pink flowers borne on long branched inflorescences attract numerous insects which come to gather nectar and pollen right beside the waves.

1. The 'path of the gods' allows one to discover the diverse flora of the Amalfi peninsula.

2. To combat erosion and to conserve the volcanic soil deposited on top of the limestone rock, crops are cultivated on tightly packed terraces descending all the way down the steepest slopes.

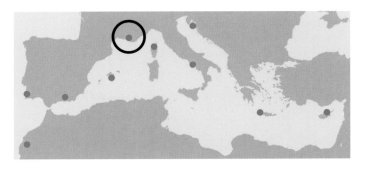

The calanques of Marseilles

When you leave the urban area of Marseilles and enter the rocky landscape of the calanques the contrast is striking. Abutting the city, this powerful limestone massif has for centuries been subject to strong pressure from human society. The presence everywhere of white rock indicates the long history of fires that have transformed the landscape. Apart from a few areas that are suitable for cultivation, the calanques have always been the domain of sheep and goats; grazing them here goes back to the first Greek colony established at Marseilles in the sixth century BCE. From then until recent times, shepherds regularly set fires to open up the cover and promote the regrowth of the grass in autumn. Today there are no longer any sheep or goats, but fires continue to sweep through the garrigue: between 1964 and 2004, the total area burnt amounted to more than the total area of the Massif des Calanques.

The frequency of fires has favoured the predominance of plants that possess the best fire-resistant strategies. On the vast screes near the Col de la Gineste, for example, the landscape has a curious leopardskin-like appearance, due to the alternation of bare rock and round patches of kermes oaks, which have a tireless ability to produce suckers after a fire. Nevertheless, in spite of the difficult conditions caused by fire, the rocky soil and the aridity increased by the force of the mistral, the Massif des Calanques has over the centuries become a remarkable 'landscape garden' of a beauty that attracts almost a million visitors every year. The complex mixture of screes, dense garrigues, flat grassy areas, rocky plateaux, shady valleys and high cliffs creates a mosaic of ecosystems that are home to an astonishing flora: almost 1000 plant species have been recorded in the Nature Reserve of the Calanques. Many species characteristic of the limestone garrigues of the South of France are found here, most of which are easy to grow in gardens. Cistus (*Cistus albidus*), autumn-flowering heather (*Erica multiflora*) and *Daphne gnidium* form low-growing expanses beneath the holm oaks. The stony ground bordering the paths is beautified by the fine clumps of hollow-stemmed asphodels (*Asphodelus fistulosus*) mingled with white sun roses (*Helianthemum apenninum*), aspic lavenders (*Lavandula latifolia*) and creeping germanders (*Teucrium aureum*). The meagre grassy spaces are home to *Brachypodium retusum*, dwarf irises (*Iris lutescens*), common thyme (*Thymus vulgaris*), succulent sedums (*Sedum sediforme* and *S. ochroleucum*) and thousands of *Aphyllanthes monspeliensis* with their bright blue flowers. The peaks most exposed to the mistral are given rhythm by cushions of santolinas (*Santolina chamaecyparissus*) and broom (*Genista lobelii*), while the screes are colonized by euphorbias (*Euphorbia characias*) and rush-like coronillas (*Coronilla juncea*).

On days when the wind blows from the sea a curious phenomenon can be observed on Mount Puget, the summit of the massif, which reaches a height of more than 500m (1640ft) above sea level. By a foehn effect, the air over the south-facing slopes becomes colder as it rises so that a dense cloud sometimes forms at the summit, suddenly reducing visibility to a few dozen metres. On the rocky peak of Mount Puget one can just make out the bright pink valerians (*Centranthus ruber*) and the pure white *Laserpitium gallicum* that decorate the expanse, like a stony desert taken over by fog. In valleys with a northern exposure the vegetation is lusher, with countless viburnum bushes (*Viburnum tinus*) growing together with buckthorns, honeysuckle (*Lonicera*), phillyreas, coronillas, jasmine, amelanchiers and bupleurums, among which are sometimes found old Montpellier maples (*Acer monspessulanum*), sumacs (*Rhus coriaria*) and magnificent whitebeams (*Sorbus aria*) covered in red berries in autumn. On the coastal fringe, cushions of yellow asteriscus daisies (*Pallenis maritima*) grow beside aromatic artemisia (*Artemisia caerulescens* subsp. *gallica*) and rosemary, hugging the ground because of the powerful salt spray. Unaffected by wind and salt, twisted old Aleppo pines have managed to survive all fires by clinging to fissures in the cliffs, suspended high above the sea.

1. Two millennia of grazing and fires have transformed the landscape of the calanques. *Erica multiflora* after its flowering is over on the coastal path between Callelongue and Sormiou.

2. The foehn effect on Mount Puget: the stony expanses resemble a flowering desert taken over by mist.

3. Leopardskin-like landscape near the Col de la Gineste: recurrent fires favour those species able to regenerate fastest after a fire, such as the kermes oak and the Aleppo pine.

4. *Erica multiflora* in full flower in autumn, with the Dévenson cliffs in the background.

5. *Coronilla juncea* colonizes peaks exposed to the mistral.

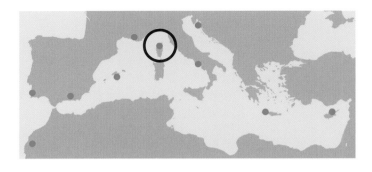

Cap Corse

The 'customs officers' path along the coast to the point of Cap Corse crosses a landscape that has been shaped by the wind. With more than 300 windy days a year, exposed in turn to the force of the tramontane, the mistral and the libeccio, the entire western side of Cap Corse has been sculpted by salt spray. The plant groups are layered according to their proximity to the sea. Nearest to the waves, directly showered in salt and making the most of the fine layers of sand and dust that accumulate in hollows in the granite rock, *Frankenia laevis* and the ground-hugging rosettes of plantains (*Plantago coronopus* subsp. *humilis*) form little green patches that are adapted to the extreme conditions. Above the line demarcating rocks that are swept by the waves during storms, the downy grey foliage of cinerarias (*Senecio cineraria*) mixes with the succulent stems of samphire (*Crithmum maritimum*) and the silvery blue of Balearic euphorbias (*Euphorbia pithyusa*). Higher up, dark flows of lentisk (*Pistacia lentiscus*) alternate with a series of aromatic cushions, regularly cut back by the salt, of *Stachys glutinosa*, white-flowered *Teucrium capitatum*, silvery *Helichrysum italicum* and a profusion of rosemaries and cistuses growing in intermingled sweeps.

A dense and often impenetrable maquis fills the small valleys where the soil is deeper. This maquis has been regularly burned in order to create grazing land and is thus composed of species able to regenerate from the rootstock immediately after a fire. The tightly packed bushes of myrtle (*Myrtus communis*), arbutuses (*Arbutus unedo*), narrow-leaved mock privet (*Phillyrea angustifolia*) and tree heathers (*Erica arborea*) form thick sheets which the salt-bearing winds that rush into the valleys shape into creases and folds.

All along the coast, the seasonal cycle of the vegetation is expressed in a landscape of changing colours. In autumn the deep blue-mauve of rosemary lights up the land. In winter the maquis of heather and arbutus is such a bright green that it might even make one think of Irish landscapes. In April the mauve flowers of lavender (*Lavandula stoechas*) mingle with the bright yellow of *Calicotome villosa*, while by the afternoons the white petals of cistuses (*Cistus salviifolius*) lie strewn on the ground, the brief life of the individual flowers over. By June, cat thyme (*Teucrium marum*) – not in fact a thyme but a germander – is smothered in mauve flowers buzzing with bees, and the helichrysums are covered in golden-yellow flower heads. At the beginning of July, among the dark foliage of the myrtles now covered in pure white buds, Montpellier cistuses form great masses of bright orange, their linear, sticky leaves now hidden by their countless dry inflorescences with sepals that retain their colour all summer.

In particularly dry years the summer colours of the maquis become as magnificent as they are unusual. In September 2007, after a total rainfall of less than 150mm (6in) during the first nine months of the year, the shining foliage of the myrtles had become a flamboyant red, the orange-red arbutuses were beginning to lose their leaves, and the rosemaries, having turned an amazing translucent yellow because of the drought, formed a golden edging to the sombre masses of junipers (*Juniperus phoenicea* subsp. *turbinata* var. *turbinata*). In such years, at the end of summer one may sometimes see the improbable sight of lean cows searching for the last tender shoots of phillyrea or heather, wending their way slowly through the sun-baked maquis as they patiently wait for rain.

6

1. The entire western side of Cap Corse has been sculpted by salt spray: dark green sheets of lentisks and myrtles, orange-coloured balls of the dry inflorescences of cistuses at the end of summer, and silver clumps of *Calicotome villosa*.

2. Well adapted to resist salt spray, *Helichrysum italicum* forms a landscape of regular-shaped balls on slopes by the coast.

3. Chiselled by wind and salt, junipers (*Juniperus phoenicea*) grow right down to the shore.

4. Capable of regrowing from the rootstock after fire, heather (*Erica arborea*) is one of the plants that dominate the landscapes of Cap Corse.

5. Between Nonza and Punta di Canelle, the slopes on the roadsides are covered in rosemary with magnificent dark blue flowers.

6. The white flowers of *Teucrium capitatum* blend into its silver foliage.

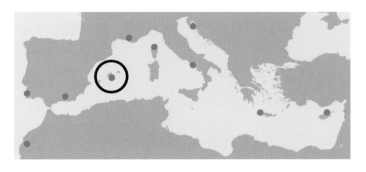

Mallorca: the steppes of the Formentor peninsula

The Formentor peninsula is a continuation of the mountain chain of the Serra de Tramuntana in north-west Mallorca. On this rocky tongue of land extending into the sea, a robust grass that is adapted to both drought and fire dominates the landscape. The clumps of *Ampelodesmos* have a double strategy to resist fire: on the one hand they resprout immediately after a fire thanks to their vigorous rootstock which survives the passage of the flames, and on the other they reseed rapidly afterwards, with flowering and seed production being particularly abundant the following year. The seeds, scattered widely by ants, germinate with the first autumn rains. Better equipped than other species to occupy the space left bare by fire, *Ampelodesmos* has progressively become dominant after centuries of regular burning. The rocky expanses of the Formentor peninsula were transformed into vast monospecific steppes, unusual among the landscapes that surround the Mediterranean – their monotonous succession of vigorous clumps of this grass called to mind rather the pampas landscapes of South America.

Today, the reconquest of the area by a greater diversity of plants is transforming the grassy steppe into a 'landscape garden' in full evolution. In late summer the golden clumps of *Ampelodesmos* create a pointillist landscape punctuated by the dark shapes of dwarf palms (*Chamaerops humilis*), Balearic boxwood (*Buxus balearica*) and wild olives (*Olea europaea* subsp. *sylvestris*). Carpets of cyclamens (*Cyclamen balearicum*) spread below shrubby St John's wort (*Hypericum balearicum*) and Balearic buckthorn (*Rhamnus ludovici-salvatoris*), as rounded as topiaries. Species less able to withstand fire have long ago taken refuge along the mountain crests or on inaccessible ledges.

The north side of the peninsula has spectacular cliffs plunging into the sea. If you lean over the parapet of the Es Colomer belvedere, hanging over the void, you will see colonies of plants exploiting the tiniest cracks in the rock: silver cushions of *Helichrysum orientale*, magnificent flowering balls of Cretan scabious (*Lomelosia cretica*) and blueish clumps of the jointed stems of *Ephedra fragilis*, thriving in their vertical home, protected from both sheep and fire, about 200m (656ft) above the sea. At the tip of the peninsula the plants show a remarkable evolutionary convergence in their survival strategies, which sometimes makes it hard to distinguish between plants belonging to different families when they are not in flower; their similar structure in the form of spiny balls allows them to better withstand salt, wind, drought and herbivores. Known locally as *coixinets de monja*, or monks' cushions, *Launaea cervicornis* with its curious branching spines like antlers, Balearic astragalus (*Astragalus balearicus*), spiny germander (*Teucrium subspinosum*) and *Dorycnium fulgurans* brave the storms as they colonize the rocks around the Formentor lighthouse.

6

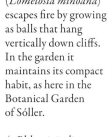

1. The golden expanses of *Ampelodesmos mauritanicus* form a steppe punctuated by dark clumps of dwarf palms (*Chamaerops humilis*), Balearic boxwood (*Buxus balearica*) and wild olives.

2. *Teucrium subspinosum* grows as a tight cushion, protected by the pedicels of its inflorescences which become as hard as spines.

3. The Cretan scabious (*Lomelosia minoana*) escapes fire by growing as balls that hang vertically down cliffs. In the garden it maintains its compact habit, as here in the Botanical Garden of Sóller.

4. *Phlomis italica* in the Serra de Tramuntana.

5. *Ephedra fragilis* on a clifftop.

6. The dark green leaves of Balearic St John's wort (*Hypericum balearicum*) are equipped with essential oil glands which protect the plant from sheep.

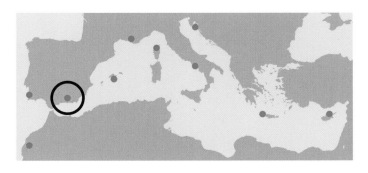

Sierra Nevada: the road to the Suspiro del Moro pass

The road to the Suspiro del Moro (Moor's Sigh) pass is a narrow, winding one that climbs up through the foothills of the Sierra Nevada. As it progresses through a succession of climatic conditions, from the hottest to the coldest, and through different types of rock, schist, marble and dolomite for tens of kilometres it offers a concentration of the diversity of southern Andalucia's flora. As one leaves the coast, the roadsides are colonized by supple bushes of silver-leaved *Retama sphaerocarpa*, laden with small pale yellow flowers. Cascading masses of fringed pinks (*Dianthus broteri*), bright green carpets of *Putoria calabrica*, clumps of fine oregano-scented lavender (*Lavandula multifida*), giant snapdragons (*Antirrhinum barrelieri*) and bright blue *Trachelium caeruleum* make astonishing vertical gardens on the rocky ledges above the road, covered in butterflies. Higher up, the limestone slopes are thick with densely packed leguminous species in different shades of yellow: golden-yellow broom (*Genista umbellata*), lemon-yellow *Ononis speciosa* and pale sulphur-yellow *Anthyllis cytisoides*.

After a few kilometres one enters the realm of cistuses, with dense populations of the silver-leaved *Cistus atriplicifolius*, bearing bright yellow flowers that are preceded by fine silky red buds. As one rounds a corner where the blue-mauve of Spanish or butterfly lavender (*Lavandula stoechas*) indicates the presence of acid soil, gum cistuses (*Cistus ladanifer*) glistening with resin make scented colonies on the dark schist screes. When once more the soil is alkaline, downy cistuses (*Cistus albidus*) mingle with sun roses (*Helianthemum syriacum*), French lavender (*Lavandula dentata*), pale mauve phlomises (*Phlomis purpurea*) and *Bupleurum gibraltaricum*, its acid-green umbels opening at the height of summer when other plants are dormant and attracting a multitude of insects. In a striking mimicry of resistance strategies, rosemary self-seeds among rosemary-leaved cistuses (*Cistus clusii*), which bear delicate white flowers that seem poised on their linear dark green leaves.

As one crosses the first pass, at about 800m (2625ft), the landscape changes abruptly. Grazed by goats and battered by the wind – hot in summer and icy in winter – the Venta del Fraile plateau is home to plants able to withstand tough conditions. The large cushions of woolly lavender (*Lavandula lanata*), with ash-grey foliage from which long spikes of mauve flowers emerge, form the structure of a remarkable aromatic garden, where mastic-scented thyme (*Thymus mastichina*) alternates with *Thymus longiflorus* with its long tubular flowers, *Thymus zygis* with verticillate flowers, and golden carpets of *Teucrium lusitanicum* subsp. *aureiforme*. Between euphorbias (*Euphorbia nicaeensis*) and asphodels (*Asphodelus macrocarpus*), Spanish sage (*Salvia lavandulifolia* subsp. *vellerea*) spreads in silvery carpets of allelopathic vegetation (see p.175) which prevents any competition from other species.

On lower slopes that have undergone recent disturbance, bright orange santolinas (*Santolina elegans*) grow among violet-blue gromwells (*Lithodora fruticosa*). A natural phlomis hybrid (*Phlomis × composita*) self-sows between its parents, the narrow-leaved *Phlomis lychnitis* and *Phlomis crinita* subsp. *malacitana*, which has large woolly inflorescences bearing bicoloured flowers, orange and brown. At the edge of the plateau, colonies of white snapdragons (*Antirrhinum hispanicum*) surround the bases of the wind turbines that mark the Suspiro del Moro pass. According to the legend, it was here that King Boabdil, the last Nasrid ruler of Granada, wept when he looked back at the distant palace of the Alhambra that he would never see again. Above the pass the summits of the Sierra Nevada remain snow-covered well into spring, while the small blue flowers of *Erinacea anthyllis* are already lighting up the slopes above the plain of Granada.

1. The silver-leaved *Cistus atriplicifolius* forms dense populations after fires.

2. Alfa grass (*Stipa tenacissima*) is a robust grass that self-seeds on rocky ground at roadsides.

3. *Ononis speciosa* is a shrub with dark, sticky and aromatic foliage from which spectacular spikes of flowers emerge in spring.

4. *Thymus longiflorus* is pollinated by butterflies which plunge their proboscises deep into the tubular flowers.

5. *Helianthemum syriacum* grows in stony soils and on limestone screes.

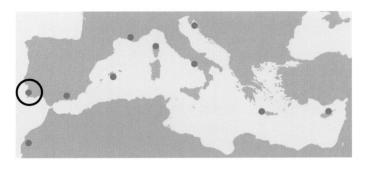

Algarve: the Costa Vicentina

With its high limestone cliffs towering above the Atlantic, Cape Saint Vincent is a windswept rocky promontory that marks the south-westernmost point of Europe. The Costa Vicentina, the coast stretching north from Cape Saint Vincent, has a surprising peculiarity: plants grow on a partially fossilized dune perched on a clifftop. At the peak of the last Ice Age, about 22,000 years ago, the sea level was approximately 130m (426ft) lower than it is now; all along the Atlantic coast of south-west Portugal the sea had retreated several dozen kilometres, leaving immense expanses of the sand which forms the seabed here. This exposed sand, carried by gusty winds, gradually created a line of dunes almost 100m (328ft) higher up, corresponding to the vortices of turbulence at the top of cliffs. Today the cliffs along this coast once more plunge directly into the sea: the stretches of sand that gave rise to the dune are now entirely submerged and the relic dune has become a geological curiosity preserved through the millennia. The perfectly drained environment of this unusual dune is home to a unique 'landscape garden' that is able to remain in a state of long-term equilibrium and bears witness to the remarkable gardening possibilities offered by poor, well-drained soil.

The Costa Vicentina is without doubt one of the best places in the Mediterranean Basin to see spectacular spring flowers. Balls of linear-leaved cistuses weighed down by yellow flowers (*Cistus calycinus*), cushions of cream-flowered germanders (*Teucrium vincentinum*) and shrubby thrift bristling with pink flowers (*Armeria pungens*) fight for a place in the interlaced carpets of white and pinkish-mauve cistuses (*Cistus salviifolius* and *C. crispus*). Pure white candytuft (*Iberis procumbens*) and gentian-blue pimpernels (*Anagallis monelli*) self-seed between prostrate *Onobrychis humilis* and purplish-red snapdragons (*Antirrhinum majus*). Camphor-scented thyme, with deep mauvey-pink flowers (*Thymus camphoratus*) grows interwoven among the bright yellow cushions of broom (*Stauracanthus spectabilis*). The elegant blue flowers of *Ginandriris sisyrinchium* and the stocky inflorescences of yellow-green *Thapsia villosa* emerge among clumps of mauve lavender (*Lavandula pedunculata*).

In this exposed landscape, the violence of the wind has permanently shaped the evergreen vegetation. The emblematic plant of this coast is a cistus with large white flowers, whose dark and surprisingly sticky leaves give off a powerful scent of ladanum (*Cistus ladanifer* var. *sulcatus*); capable of surviving in the first line of salt spray, it grows in compact balls right up to the cliff edges that tower over the sea. Surrounded by carpets of juniper (*Juniperus phoenicea* subsp. *turbinata*) flattened by the wind and by enormous rounded masses of white-fruited *Corema album*, these cistuses form the main structure of a magnificent vegetation consisting of numerous ball- or cushion-shaped species. The result of an unusual geological history and subject to tough conditions, the Costa Vicentina is like a model Mediterranean garden; it has the double advantage of being richly endowed with flowers in spring and remaining attractive at other times of year, including during the dry period in summer, by virtue of the undulating rhythm of its evergreen vegetation sculpted by the wind and salt spray.

1. An extraordinary flowering landscape on dunes on the west coast of the Algarve: *Armeria pungens, Stauracanthus genistoides, Corema album, Cistus calycinus* and *Helichrysum italicum.*

2. The palhinae cistus (*Cistus ladanifer* var. *sulcatus*), whose leaves give off the powerful incense-like scent of ladanum, grows on top of the limestone cliffs of Cape Saint Vincent.

3. *Thymus camphoratus* and *Onobrychis humilis* grow in the stony or sandy soils behind the dunes.

4. *Ginandriris sisyrinchium* is a member of the iris family which spreads on the dunes in spaces between the shrubs.

5. *Pallenis maritima* establishes itself on poor, rocky land directly exposed to salt spray.

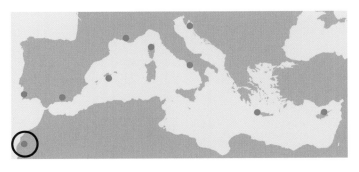

Crossing the Atlas Mountains: the road to Tizi N'Test

The chergui is a strong wind that blows from the Sahara Desert. When it meets the high barrier of the Atlas, it fills the valleys, blusters along the south-facing slopes of the mountain, crosses its crest and descends, by now completely dry and even hotter, into the plains of Marrakech. When the chergui blows in summer the temperature in Marrakech rapidly becomes scorching, often exceeding 45°C (113°F) and sometimes even touching 50°C (122F°), as in the exceptional heatwave of July 2012. These climatic extremes make life difficult for humans and plants alike. The traditional architecture of the Berber villages in the valleys of the Atlas, where the mud-built houses blend into the landscape, are designed to control the temperature within thanks to the thermal inertia of their thick clay walls. In the foothills of the Atlas, garrigue plants also rely on multiple strategies to withstand heat and long periods of drought. All along the valley that leads to the village of Imlil, the point of departure for climbing to the summit of Toubkal, dwarf palms or *doum* (*Chamaerops humilis* var. *argentea*) take on an amazing metallic colour: the silvery-blue waxy cuticle covering the leaves enables the plant to resist heat and intense sunlight.

Near the souk of Asni, on the road that crosses the Atlas and goes to Taroudant, the yellowy-green flowers of *Anagyris foetida*, the cream-coloured flowers of *Chamaecytisus mollis* and the blue or mauve flowers of *Teucrium fruticans* create a colourful garden in the middle of winter. These various species, which come into flower as soon as water is available to them after the autumn rains, adopt a common strategy to get through the summer: they reduce their leaf surfaces in order to survive during the hot period by slowing down photosynthesis.

As one follows the valley of the river N'Fiss and climbs the mountains in the direction of the pass of Tizi N'Test, the landscape changes rapidly. Near the potters' village of Ijoukak, on the red clay used to make tagines, Barbary thujas (*Tetraclinis articulata*) cling to the slopes; these are full of gullies as a result of the erosion caused by the rare but torrential rainstorms which open up deep clefts in the clay slopes denuded by grazing. Taking refuge on the steep slope above the road where they are safe from sheep, midnight-blue lavenders (*Lavandula maroccana*) grow among the curious orange-flowered *Linaria ventricosa*, bright yellow coronillas (*Coronilla ramosissima*), brilliant pink valerians (*Centranthus longiflorus*) and spectacular polygala bushes, with leafless glaucous stalks covered in late winter with mauve and yellow flowers (*Polygala balansae*). Below, white *Retama monosperma* dominates the wadi, which is bordered by tamarisks and oleanders. Higher up, the Barbary thujas give way to an open forest of cypresses (*Cupressus atlantica*), the relics of what was once a much more extensive ancient forest. A rich layer of aromatic sub-shrubs has colonized the ground beneath the trees because the upright habit and weeping foliage of the Atlas cypresses allow light to reach the soil. The scents of a rue with finely cut leaves (*Ruta montana*), of white artemisia (*Artemisia herba-alba*), of grey lavender (*Lavandula dentata* var. *candicans*) and of magnificent savory-leaved thyme (*Thymus saturejoides*) mingle with the strong aroma of cypress wood, perfuming the air in the entire valley.

As you go higher, villages become rare, the last mud houses being huddled by almond trees grown on terraces whose curves follow the sinuous course of the wadi. Near the mountain peaks, laurel-leaved cistuses (*Cistus laurifolius* subsp. *atlanticus*) and thick-leaved globularias (*Globularia alypum*) self-seed among the evergreen oaks (*Quercus rotundifolia*). When you arrive at the Tizi N'Test pass, you suddenly see, lying some 1700m (5577ft) below you, the immense stretch of the Sous valley, flooded with light and punctuated by thousands of dark green argan trees. Above the pass, on schist screes marked by the massive silhouettes of ancient junipers (*Juniperus thurifera*), spiny cushions of *Ptilotrichum spinosum* and *Bupleurum spinosum* seem able to resist anything, buried under snow in winter then blasted by heat in summer.

1. An Atlas cypress by the N'Fiss wadi on the road between Marrakech and Taroudant.

2. *Linaria ventricosa*, with silver-blue foliage, grows at an altitude of 2000m (6562ft) south of the Tizi N'Test pass.

3. The inflorescence of *Linaria ventricosa*.

4. Colonizing the ground beneath Atlas cypresses, the savory-leaved thyme (*Thymus saturejoides*) flowers spectacularly in early summer.

5. *Polygala balansae* has glaucous stems bare of leaves, which enables the plant to limit water loss in periods of extreme heat or drought.

LEARNING FROM THE LANDSCAPE: THE GARDEN AS AN ECOSYSTEM

Stone, sun, wind and drought: garrigue plants have evolved to cope with such tough conditions by multiplying their strategies to withstand them. These plants have specialized in order to thrive where most other species would be incapable of surviving at all. The well-meaning gardener who attempts to grow lavenders, santolinas, rosemaries or cistuses in soil enriched with potting compost and carefully watered by a drip system has taken a wrong path, for the majority of garrigue plants will not be able to tolerate the horticultural conditions of a traditional garden. In their natural habitat, lavenders are happy on the windswept mountains of Provence, santolinas seed themselves among the cracks in the rocks of the calanques of Marseilles and rosemaries colonize rocky slopes on the Corsican coast, while cistuses, which have the evocative common name of rock roses, adorn the stoniest and poorest soils with their crumpled flowers all around the Mediterranean, from the mountains of Lebanon to the valleys of the Anti-Atlas. Consequently, adapted as they are to a frugal environment, most garrigue plants have difficulty acclimatizing to a garden where the soil is too rich, where they receive regular irrigation and where dampness prevails in winter. To welcome them into our gardens, therefore, we must offer them the particular conditions that correspond to what they like: soil with perfect drainage, as poor as possible.

The gardener who is prepared to abandon traditional horticultural techniques launches himself or herself into an exciting adventure. Improving drainage and impoverishing the soil enables one to grow a magnificent range of drought-resistant plants, still little known because they are rarely cultivated in the gardens. Changing our growing conditions has another advantage: it enables us to set in motion a remarkable dynamic in the garden. By creating conditions close to those of their natural habitat, we encourage garrigue plants to self-seed freely. In this way the garden begins a cycle of evolution which will profoundly change our perspectives on planning and maintaining a garden. Shaped by human interventions, garrigues have always been landscapes that change, with an evolutionary trajectory that varies from place to place according to the forces that influence the landscape locally. Garrigue plants are perfectly adapted to these conditions. In the wild they take advantage of the evolution of their environment by multiplying to colonize new spaces as soon as conditions are favourable, and in the garden, when they find conditions that suit them, they express in the same way the dynamic capacity that they have acquired over millennia. Native to landscapes which are in a perpetual state of transformation, they are prime material for a garden that evolves under the gardener's guidance, inspired by the behaviour of plants in the wild.

Top: *Euphorbia characias* subsp. *wulfenii* and *Achillea coarctata*. Coming from landscapes that are in a state of constant transformation, garrigue plants are prime material for gardeners who embrace change and enjoy managing evolutionary shifts, drawing inspiration from the way the plants behave in the wild.

Bottom: Vegetation re-establishing itself after a fire on the Punta Campanella, south of Naples. The way in which garrigue plants respond to fire is one of the factors that explains the dynamism of pioneer plants in gardens.

• *Thinking about fire*

The behaviour of plants when they are subjected to fire in the garrigue is one of the factors that explain their greater or lesser dynamism in the garden. Used by humans for about 7000 years as a means of opening up land for grazing, fire has had a profound influence on Mediterranan ecosystems. Recurrent burning has a selective effect on the composition of the landscape: it favours pyrophyte species (from the Greek *pur*, fire, and *phuton*, plant), which have developed adaptive strategies that enable them to cope with fire. There are two such major strategies which allow these plants to use fire to become locally dominant. Some pyrophyte plants are killed by fire but nevertheless profit from it twice over, for the passage of the fire triggers the germination of their seeds, while at the same time creating the open landscape that they need if they are to self-seed profusely. Others are by contrast able to survive fire: they lose their leaves for only a brief period after being burnt, then regrow vigorously from new shoots sprouting from their rootstock.

Most of the sub-shrubs that form the structure of low-growing garrigues around the Mediterranean are heliophiles (from the Greek *helios*, sun, and *philein*, to love) which are happy in stony, open and sunny environments. After a fire the garrigue evolves through various stages, leading, when soil and climate conditions allow, towards a complete closure of the landscape. This pre-forest stage, which may appear after about 15 years,

A post-fire landscape in Andalucia: *Phlomis purpurea*, *Hyparrhenia hirta*, *Genista umbellata* and *Ferula communis* grow on slopes which a few years earlier had been cleared by fire.

consists of evergreen vegetation which grows progressively taller to reach a height of several metres. It creates a dark and dense environment, clearly less favourable to heliophile plants, which generally manage to survive for a few more years in semi-shade before disappearing once and for all when they are overwhelmed by competition from the layer of taller plants. Cistuses, heliophile plants characteristic of garrigues shaped by fire, have a lifespan of about 15 years in the natural conditions of poor, well-drained soil that suit them best. This lifespan corresponds to the length of time that the garrigue remains open after a fire. For heliophile plants such as cistuses there is no point in living longer, since the closing of the landscape makes conditions unfavourable to them. However, thanks to their particular manner of reproduction, these plants are able to maintain a long-term presence in the landscape.

Many heliophile garrigue plants that are linked to fire ecology have developed a remarkable survival strategy by means of their seeds. These develop in cycles and are capable of awaiting the return of favourable conditions created by fire. Every year these species produce a very large quantity of seeds; a mature cistus, for example, produces between 5,000 and 10,000 seeds annually. When they fall to the ground around the plant, the seeds form an abundant reserve that slowly accumulates in the soil, year after year. Cistus seeds are capable of retaining their ability to germinate for a long time as a result of a dormancy mechanism which enables them to await the arrival of the next fire, while simultaneously preventing the seeds from germinating in normal times. It thus effectively ensures the continuation of the species in their environment. For a cistus there's no hurry; in the cyclical evolution of the garrigue all that is necessary is to wait patiently for the flames to return, for it is fire that will create the most favourable conditions for the plant to prosper once again. On the one hand, fire opens up the landscape quite brutally, creating the sunny spaces that cistuses need; on the other hand, it abruptly breaks the seeds' dormancy, allowing them to germinate very rapidly.

Cistus seeds are protected by a hard and impermeable outer coating which prevents water from penetrating and inhibits the germination of the embryo. They can germinate only when this outer coating cracks in the heat of a fire. Depending on the species and the length of time it is exposed to heat, the dormancy of cistus seeds is broken at 50–130°C (122–266°F), which corresponds to the temperatures generally reached in the top few centimetres of soil when fire sweeps over a Mediterranean ecosystem. Thus, by opening up the space and triggering germination, fire suddenly and spectacularly favours the propagation of cistuses; after a forest fire, one may sometimes see the phenomenon of mass germination, with up to 800 cistus seedlings per 1sq m (1¼sq yd), giving the effect for a few months after the autumn rains of an astonishing carpet of tiny plantlets entirely covering the soil.

This phenomenon of mass germination, however, never occurs in gardens. In the absence of fire, the germination rate of the seeds of pyrophytes such as cistuses, shrubby sages and lavenders is generally very low, if not zero. Nevertheless, some seeds manage to germinate when the temperature of stony soil rises high enough in periods of intense sunshine to crack the outer coating of seeds lying on the surface of the soil. In our own garden we have noticed that several cistus species self-seed in areas where the stony soil is fully exposed to the sun – for example where the ground is covered with an inorganic mulch (pebbles or gravel). Our gravel garden, which reproduces an open environment in which the plants are separated by spaces covered with a layer of gravel 6–8cm (15–20in) thick, is thus home every year to many self-seeded heliophile pyrophytes – cistuses, sages, thymes and lavenders – which, in the absence of the fire they need in order to propagate themselves to the fullest extent, benefit from

the heat of the gravel in midsummer to break the dormancy of at least some of their seeds.

Drawing inspiration from the ecology of fire, the gardener can thus modulate the self-seeding of the sub-shrubs that form the core of the garrigue garden: in areas of the garden where the soil is covered with an inorganic mulch, heliophile pyrophytes self-seed more than in other parts of the garden. This self-seeding in a gravel garden has a double advantage. It ensures the long-term presence of these plants without the gardener having to replant when they come to the end of their natural lives, which generally span 10–20 years depending on the species. It also ensures that new little plants are available every year free of charge, which the gardener may choose to keep or to discard depending on the way he or she wants the different areas of the garden to develop.

• Lignotubers – a strategy for dominating the landscape

Some garrigue plants regrow even if one cuts them right down to the ground: these are pyrophyte species that can completely regenerate after fires. This is the case with many sclerophyllous plants (from the Greek *skleros*, hard, and *phullon*, leaf), which withstand drought thanks to their thick, leathery evergreen leaves. Shrubs such as myrtles, lentisks, phillyreas, viburnums, box and buckthorn, or trees such as arbutuses, olive trees, holm oaks and kermes oaks, have the ability to regrow from their lignotubers when the parts of the plant above ground have been utterly destroyed. A lignotuber (from the Latin *lignum*, wood, and *tuber*, tumour, excrescence) is a swelling rich in starches and full of dormant buds formed at the base of the trunk or on the roots of some pyrophyte

From left to right:

The dormancy of cistus seeds is broken by the thermal shock caused by the passage of fire.

Cistuses self-seeding prolifically in an area of our garden where the soil is covered by an inorganic mulch.

Lomelosia minoana germinating between paving stones in our garden. The abundance of self-sown seedlings in the garden means that every year new plants are available free of charge.

Thanks to its lignotuber, the Cyprus arbutus (*Arbutus andrachne*) is able to put out new growth from the rootstock when its trunk is destroyed by fire.

species. A host of young shoots sprout from these dormant buds when the foliage has been brutally destroyed by fire or after heavy pruning. The dense copses of holm oaks (*Quercus ilex*) which one often sees in the South of France are the result of repeated cutting by charcoal burners over the centuries, just as the hillsides covered only in kermes oaks (*Q. coccifera*), whose tangled branches grow directly from the soil, are an indication of recurrent fires that have progressively allowed the kermes oak to become the dominant species on the scale of an entire landscape.

Lignotubers indeed offer an important competitive advantage in garrigues that burn regularly. In spite of the temporary absence of photosynthesis, species that resprout from the rootstock regenerate very quickly thanks to the reserves contained in their lignotuber. Exploiting an already extensive root system capable of drawing water and mineral elements from a significant volume of soil, the foliage of plants with a lignotuber thus grows very much faster than that of a new plant which has grown from seed, whose young roots have access to water in only a limited volume of soil. In this way sclerophyllous trees and shrubs that resprout from the rootstock overtake other plants and are particularly successful in occupying the space cleared by fire; after a few years they are once more such vigorous plants that the fire-stricken landscape is quickly forgotten.

This ability to regrow from the rootstock enables plants to withstand not only fire but also many other distur-

Left: After a fire the kermes oak (*Quercus coccifera*) starts regrowing directly from its roots.

Right: In a burnt landscape the light green and bronze colours that relieve the blackened land come from thousands of newly sprouting kermes oaks which take advantage of the fire to start dominating the landscape immediately.

bances or climatic events that can destroy top growth: exceptional cold, drought, wind, salt spray, repeated cutting and grazing. The pressure of grazing on the landscapes of the Eastern Mediterranean thus often transforms plants with lignotubers into amazing sculptures as the plants find a balance between their continual regenerative power and the continual action of sheep or goats which prune them thoroughly as they browse on them day after day. Some sclerophyllous shrubs can live for decades with their branches hugging the ground as a result of grazing or climate conditions, their normal upright habit having turned into a carpeting, ground-covering form. This can be seen in lentisks, which in coastal areas subject to strong winds often form large, round sheets of dark green only tens of centimetres high, sometimes cut down by fire and then flattened by the salt borne on the powerful spindrift. Even more spectacularly, wild olives cut down by fire and then browsed on daily by goats are able to regrow continually thanks to their lignotubers. In these circumstances they may be transformed into flat carpets, their interlaced branches hugging the ground so that after many years they form a tough weft of tiny leaves that looks almost mineral, very different from the trees they would have become without the cumulative action of fire and grazing.

In the garden, sclerophyllous plants with lignotubers such as lentisks, myrtles, arbutuses, phillyreas and olives have a quality that derives from their adaptation to fire in garrigues: they live for a remarkably long time. In gardens in Corsica one may sometimes see hundred-year-old lentisks or myrtles with thick lower branches growing horizontally, allowing them to serve as rustic benches shaded by the fragrant foliage above. In the botanic garden of the Villa Thuret at Antibes, the impressive trunks of old arbutuses such as the Cyprus arbutus (*Arbutus andrachne*) and its hybrids (*A.* × *andrachnoides*, *A.* × *thuretiana*), from which ornamental bark peels in strips like that of some eucalypts, are one of the attractions in this garden renowned for its exceptional trees. In the Jardin des Plantes at Montpellier, a phillyrea (*Phillyrea latifolia*) famous for its extraordinary furrowed trunk must date from the 17th century when the garden was established. The olive tree that dominates the square in the village of Vouves in Crete, whose twisted trunk has a circumference of 12m (39½ft), is probably more than 2000 years old. Pruned by generation after generation over the centuries, this monumental olive still produces fruit and faithfully puts out vigorous new growth that regenerates its vegetation each year.

A long way from the romantic images evoked by the magnificent olives of Crete, Delphi, Mallorca or Sardinia that are thousands of years old, the pavements of the bustling streets leading into Syntagma Square in the heart of Athens are lined with low hedges of olive, clipped with

Top: Kermes oaks in the White Mountains in Crete. The pressure of grazing often transforms plants with lignotubers into amazing vegetable sculptures, the plants finding an equilibrium between their natural regenerative force and the action of the sheep or goats which prune them by browsing on them day after day.

Bottom: In the wild as in gardens, sclerophyllous plants tolerate regular pruning perfectly well. Here, garrigue plants – lentisk, phillyrea, juniper and buckthorn – are grown in pots in the Sóller Botanical Garden in Mallorca.

Opposite: Lentisks clipped to give structure to the space in the Park of Saleccia near L'Île-Rousse in Corsica. The ability of sclerophyllous plants with lignotubers to tolerate different clipping regimes, resulting from their adaptation to fire in the wild, makes them key elements in the permanent structure of a garrigue garden.

perfect regularity to a height of about 60cm (2ft), the purpose of which is no doubt to offer psychological protection to tourists from the smell of exhaust fumes and the honking of taxis. Apart from the benefit of their exceptionally long life, olives, like all trees and shrubs with a lignotuber, are perfectly tolerant of clipping in parks or gardens. Box and myrtle are ideal material for clipped edging, while narrow-leaved mock privet (*Phillyrea angustifolia*), lentisk and viburnums lend themselves to topiary of all shapes and sizes: they always regrow, whether after regular light pruning or after an occasional much more severe cutting back.

Their longevity and their ability to tolerate different pruning regimes thus make sclerophyllous plants with lignotubers, which arose from an adaptation to fire, key elements in the permanent evergreen structure of a garrigue garden. Their regeneration from the rootstock is so effective at ensuring their long-term survival that they have no need to self-seed on the spot. To avoid finding themselves in competition with their own offspring, they have adopted a

manner of propagation that allows them to colonize territory at a distance from the mother plant. Lentisk seeds, for example, germinate only when fresh. Yet a fresh lentisk seed falling on to the ground beneath the mother plant cannot germinate because the pulp surrounding it contains substances that inhibit germination; in order for it to germinate, the seed needs to be completely free of this pulp. The fleshy berries of sclerophyllous shrubs with lignotubers

such as myrtle and lentisk are very attractive to birds which eat them as soon as they are ripe, and passage through the birds' digestive system rids the seeds of their germination-inhibiting pulp. The birds then expel the seeds in their droppings, having in the meantime flown some distance away, maybe a few dozen to a few hundred metres. Sclerophyllous garrigue plants thus nearly always seed themselves beneath perching sites at some distance from the mother plant – usually upright shrubs or trees where the birds can perch safely, under which pulp-free seeds, ready to germinate, accumulate in their droppings.

Apart from their robust nature and lengthy lifespan, sclerophyllous shrubs have another great advantage in gardens that are shaded by trees, for example oaks or pines. Because of their manner of propagation, they are adapted to cope with competition from the plants that provide the perching sites beneath which they are able to germinate; thus not only can they tolerate shade and root competition, they have also developed a reduced sensivity to the allelopathic properties of trees that limit the germination of other species (see p.175). In our own garden, the ground beneath Aleppo pines is being progressively

Top: The fleshy berries of the myrtle are very attractive to birds, which disperse the seeds in their droppings.

Bottom: The garden of Rosie and Rob Peddle in Portugal. In this garden, which is wild in parts, the ground beneath old olive trees is colonized naturally by viburnum, its seeds spread in the droppings of the birds which perch on the branches of the olives.

colonized naturally by lentisks and viburnums, which we shape by light pruning to form ornamental layers of different heights. In old Mediterranean gardens that are shaded by trees, sclerophyllous plants are particularly useful for giving structure to the space with their handsome evergreen foliage, creating an underlayer reminiscent of a garrigue that is in the course of becoming closed. These plants, moreover, are often the last option possible beneath pine trees, where shade, the abundance of fallen pine needles and the dryness intensified by strong root competition create a very inhospitable environment for most other species.

The vigorous seeding of sclerophyllous plants producing seeds that are dispersed by birds can sometimes be impressive. In the Montpellier region, the old tree-filled gardens around grand old houses which at some point in their history have been abandoned are often overrun by dense viburnums, buckthorn and phillyrea forming well-nigh impenetrable copses in the midst of old box edging that has run free, giving a natural charm to these gardens that were once formal in the French style. In the different zones of a garrigue garden, colonization by

sclerophyllous plants can to some extent be decided in advance. The seeding of them is dependent upon whether or not there are plants in which birds may perch, dispersing seeds in their droppings. For example, in the open areas of our garden, which consist of plants of no great height, a flowering steppe, a dry lawn, a green terrace and a gravel garden, there is no spontaneous seeding of sclerophyllous plants, whereas it can be abundant in areas where the plants are taller, including trees and shrubs in which birds perch during their regular comings and goings. When one is planning a garrigue garden, layers of plants of different heights and an alternation of open and closed areas allow one to steer the garden's evolution by making the seeding of sclerophyllous plants either possible or unlikely.

● *Thinking about human activity*

Garrigues in the Mediterranean region are derived from a long history of human activity affecting nature. Some plants have become specialized to colonize land that has recently been disturbed during the different stages in the transformation of garrigue landscapes. The creation of paths through the vegetation, the regular passage of flocks of sheep or goats, the clearing and subsequent abandonment of cultivatable land gained for a while from the garrigue, the levelling of terraces on the hillsides and the ever-renewed construction of the extraordinary network of dry-stone walls common to landscapes around the Mediterranean – all these modifications of the environment represent many opportunities for ruderal species (from the

When planning a garrigue garden, try establishing layers of plants of different heights in order to manage the dynamics of self-seeding: heliophile shrubs will self-sow in open areas, while sclerophyllous shrubs will self-sow beneath the trees that birds perch in.

Centranthus ruber growing in scree on the side of a road. Ruderal species from the edges of garrigues are valuable to gardeners for their long flowering seasons and brightly coloured flowers.

Latin *rudus, ruderis*, meaning rubble) that have evolved to exploit land disturbed by humankind. For these opportunistic plants, the creation of a new garden is just another form of human action on nature. The passage of machinery to level the land after a house has been built, the creation of paths, walls, terraces and steps, the decompaction of the soil before planting or the making of raised beds to aid drainage: all the work that goes on in a Mediterranean garden creates ideal conditions for ruderal garrigue species, able to exploit in gardens disturbances that are similar to those with which they are familiar in the wild.

A disturbed environment is one which evolves rapidly and in which competition is acute. A ruderal plant that reproduces by seed has better chances of propagating itself if it germinates easily and if efficient pollination, for example by having flowers that are very attractive to insects, enables it to produce a large number of seeds. Ruderal species from the edge of garrigues are thus often appreciated by gardeners for their long flowering periods and brightly coloured blooms; bright colours attract pollinators, which are in return rewarded by the abundance of pollen and nectar that the flower offers, and a long flowering period increases the chances of seed dispersal. Valerian (*Centranthus ruber*), for instance, has a first flowering from May to July then another from September to November and in mild climates sometimes flowers all year round. Its pink, bright red, mauve or white flowers, rich in nectar, are visited by a ceaseless ballet of bees, bumblebees and butterflies over the course of the seasons, thus guaranteeing pollination throughout the whole flowering period. One may for example often see the hummingbird hawk-moth, recognizable by its hovering flight, plunging its proboscis into the narrow spur at the end of each valerian flower to suck nectar from it (*centranthus* comes from the Greek *kentron*, spur, and *anthos*, flower, referring to the shape of the nectar-bearing part of the plant). The light seeds of valerian are equipped with feathery appendages and are dispersed by the wind over long distances. They germinate easily in soil that has recently been dug,

between stones or in the crannies in dry-stone walls, and in a few years valerian can spontaneously form very decorative colonies in the garden.

The flowering of herbaceous ruderal species is spread over almost the whole year. Pitch trefoil (*Bituminaria bituminosa*), fennel (*Foeniculum vulgare*), mallow (*Malva sylvestris*) and wild chicory (*Cichorium intybus*) adorn the edges of garrigue paths in summer with their mauve, yellow, pink and clear blue flowers. Calaminth (*Calamintha nepeta*) is covered in butterflies which come to sip nectar from its narrow blue flowers in late summer. The bright yellow flower heads of sticky fleabane (*Dittrichia viscosa*) open in autumn, while hollow-stemmed asphodel (*Asphodelus fistulosus*) begins to flower in winter, often from mid-February. Sweet scabious (*Scabiosa atropurpurea*), wild carrot (*Daucus carota*), smooth golden fleece (*Urospermum dalechampii*) and salsify (*Tragopogon porrifolius*) flower all through spring, attracting a multitude of insects which come to take advantage of their nectar and pollen.

Seeds have extremely varied strategies in order to better multiply on disturbed ground. For example, the small black seeds of hollow-stemmed asphodel are contained in fruits in the form of open-topped capsules arranged along upright stems which harden as they dry out after flowering; these stiff stems, shaken in every direction by the

wind, project the seeds in the manner of a catapult, with the result that the plant is propagated year after year by prolific self-seeding around the mother plant. Salsify seeds have a pappus on top of them, a downy appendage in the form of a parachute, which enables the seeds to float on the wind, borne on rising thermals. The light seeds of sticky fleabane are also dispersed over large distances: this is a pioneer plant which seems to appear from nowhere as soon as an empty spot in the garden is available to its seeds, carried on the wind throughout the winter.

In order to colonize rapidly spaces left free by recent disturbances of the environment, some garrigue sub-shrubs, such as *Dorycnium*, euphorbias, ballotas, phlomises, coronillas and bupleurums, adopt the same strategies as herbaceous ruderal plants: flowers that are attractive to pollinators, a large number of seeds, efficient dispersal and rapid germination. These species behave like pioneer plants just as much in gardens as in the wild. For instance, euphorbias (*Euphorbia characias, E. rigida, E. ceratocarpa* as well as numerous other species) self-seed along paths or in grazed

garrigues, the irritating latex that oozes out whenever a part of the plant is injured protecting them from sheep and goats. In the garden, euphorbias are exceptionally effective colonizing plants: their long flowering season and the abundance of nectar produced by the large shiny glands on their inflorescences, which attract countless pollinators, allow them to produce a very large number of seeds. Their dispersal could be called ballistic – when the fruits are ripe

From left to right: Calaminth (*Calamintha nepeta*) is covered in butterflies sipping nectar from its flowers in summer and autumn. Its long flowering period ensures that it produces a large number of seeds which germinate with the first rains.

The seeds of salsify (*Tragopogon porrifolius*) are topped with a downy appendage like a parachute (pappus) which enables them to spread far, floating on the wind at the mercy of rising thermals.

Dittrichia viscosa is a pioneer plant whose seeds are carried by the wind; it rapidly colonizes ground that has recently been disturbed.

the seeds are projected a few metres away from the mother plant as their capsules explode in the heat, making a characteristic sharp sound. A secondary dispersal of euphorbia seeds is subsequently carried out over a longer distance by ants, which are attracted by the elaiosomes (from the Greek *elaion*, oil, and *soma*, body), fleshy structures on the seeds that contain a substance rich in lipids; ants love these and criss-cross the garden in search of euphorbia seeds to carry back to their nests as food for their larvae. Since the elaiosomes are the only bit they want, they discard the seeds all around their anthills where the aerated soil, softened by the mixture of their excrement and the bodies of dead ants regularly cleaned out from the nest, provides ideal conditions for germination.

Coronilla (*Coronilla valentina* subsp. *glauca*) is the first shrub to flower in winter in our garden, usually at the end of January. The flowers are deliciously scented and insects can spot their bright yellow corollas from afar in the winter landscape. Coronilla produces abundant seeds: in summer, when the plant loses some of its leaves in order to resist drought better, it is laden with delicate articulated seedpods that give a striking effect. When they fall to the ground these seedpods break into short oval segments, each containing a single seed. The segments are inflated with air, so they float well and provide the seeds with a light craft, like a miniature canoe, to carry them away from the mother plant; the seeds use the heavy rains of autumn as a means to migrate in the garden, sailing along

on puddles or on the tiniest of temporary rivulets. When the rain stops, the seeds amass at the edges of paths or at the foot of any obstacle that has acted as a dam and blocked their way. If they have enough light, they germinate rapidly: seedlings only a few centimetres high can form a continuous line all along gravel paths and in places free of vegetation. In our garden, *Coronilla valentina* subsp. *glauca* serves as a fill-in plant: it occupies gaps in our plantings, replacing old plants that have reached the end of their natural life spans.

Shrubby hare's-ear (*Bupleurum fruticosum*) has a flowering period which is the opposite of that of coronillas. In order to attract pollinators better it flowers right in the middle of the most difficult season, at the height of the dry period, when the majority of woody garrigue plants have long been dormant: its summer flowering makes it unbeatable in garrigue gardens. In July and August its umbels of brilliant yellow flowers, glistening with nectar, are covered in insects which often find here one of the last sources of food to see them through the summer. Efficient pollination allows the plant to produce an extremely large number of seeds. A mature bupleurum can have almost 1000 inflorescences in spreading umbels, making a magnificent sight in the summer garden, and each umbel produces more than 200 seeds, so that every autumn something like 200,000 seeds rain down on the ground around the mother plant. As with most plants that act as pioneers on disturbed land, bupleurum seeds do not have

A garden colonized by self-sown euphorbias: *Euphorbia rigida* in the foreground and *Euphorbia characias* subsp. *wulfenii* in the background.

any dormancy strategy – fresh seeds germinate very rapidly provided they have enough light. They germinate if the soil is bare of all vegetation; in a few years the tightly packed young plants will form a dense mass, covered in its turn with flowers and then seeds, enabling the plant to colonize slowly but surely any empty spaces in the garden.

• *Thinking about stone*

The recurring cycle of human disturbances has had a major consequence on the soil of garrigues. The triad *ager-saltus-sylva* (farming, transition, forest) governed the organization of space in traditional Mediterranean landscapes for a long time. The plains and hills nearest to villages were devoted to growing food crops, while less accessible slopes were home to grazing land or forests. Grazing lands were set on fire regularly to open up the vegetation and trees were cut to provide wood for construction and heating, favouring those trees able to regrow as coppices after being cut to the ground, such as the holm oak. Cutting down the vegetation, clearing the land by burning and the pressure of grazing led, by repeatedly leaving the soil bare, to an inexorable process of erosion. Run-off down hillsides after the sometimes heavy rainstorms characteristic of the wet season in Mediterranean climates carried away the soil little by little until the bedrock was exposed. In this way Mediterranean hillsides gradually evolved into the stony landscapes typical of

Top: *Coronilla valentina* subsp. *glauca* self-seeds abundantly into empty spaces in the garden, filling gaps in the beds.

Bottom: Glistening with nectar, the flowers of *Bupleurum fruticosum* are covered in insects making the most of one of the last sources of food that will help them get through the summer. Thanks to efficient pollination by insects, a mature bupleurum can produce up to 200,000 seeds in autumn.

garrigues as we know them today. This slow evolution over the course of millennia has favoured the development of an amazing flora of specialized species, for it is on stony, poor and well-drained land that most garrigue plants are happy.

The rocky or stony landscapes that we see all around the Mediterranean tell of the extent of the erosion which has removed the topsoil in garrigues. Indeed, the erosion process can sometimes still be seen in action today, for example after fires in the garrigues or in areas where over-grazing has exposed the soil directly to the elements. In August 1989, a spectacular fire broke out on Mont Sainte-Victoire, near Aix-en-Provence. Helped by the drought and the strength of the mistral, within the space of a few days this exceptionally fierce fire transformed the famous Provençal landscape into a long, sombre, charred expanse, visible for kilometres around. The abundant rains of the following autumn and spring hollowed out thousands of micro-watercourses in the outcrops of marl

on the southern slopes of Mont Sainte-Victoire; the silt and sand carried away during the first year formed a thick layer in the alluvial zones of the valley floor below. Similar erosive phenomena can be caused by repeated cutting down of the vegetation or by over-grazing, such as the large-scale transformation of the landscape currently occurring in southern Morocco, where some of the valleys

in the Atlas are undergoing serious erosion under the combined action of tree felling and extreme pressure from grazing. The denuded slopes are furrowed by a huge network of gullies becoming deeper with every passing year, carrying the soil away during the rare but often heavy rains in this mountain landscape.

To the north of Balagne in Corsica, the Agriates desert is a recent example of the way a landscape is altered following erosion. For centuries the Agriates region was famous as one of the breadbaskets of Corsica; farmers came here seasonally from neighbouring regions to cultivate wheat, citrus fruit, olive trees and figs, while the areas of maquis between the fields were used as winter pastures by shepherds when they brought their flocks down from the mountains. The repeated clearing of the maquis by fire in order to foster grazing progressively led to soil erosion and the subsequent abandonment of the fields. The Agriates region, a land of plenty until the 19th century, has now been invaded by a sparse maquis dominated by pyrophytes which recall a history of fires: in late winter, between the myrtles and the tree heathers, among the fresh new leaves of asphodels and giant fennels, spreading bushes of rosemary partially cover the granite outcrops, their flowers a particularly intense blue.

In the islands off the Adriatic coast in northern Croatia, erosion dates back much further. The long history of fires, grazing and wood-cutting, exacerbated by the pressure on the vegetation by the salt spray borne on the bora or the jugo, the prevailing winds that sweep over the coasts of Istria, has laid bare the extraordinary karstic relief of the mother rock. White limestone is the dominant element in this spectacular rocky landscape, where sage (*Salvia officinalis*), thyme (*Thymus longicaulis*), horehound (*Marrubium vulgare*) and globularia (*Globularia meridionalis*) form small clumps between the stones. The same type of erosion of karst is sometimes seen in other climate zones. On the west coast of Ireland, the Burren and the Aran Islands are famous for their landscape of limestone

The triad *ager-saltus-sylva* (agriculture-transition-forest) has long decided the organization of space in traditional Mediterranean landscapes.

From top to bottom: Vines growing among stones near Split, Croatia.

Landscape grazed by goats on the Rodopou peninsula, Crete.

Coppiced holm oaks resulting from cutting wood for charcoal burning. Forest on Mount Liausson, above Mourèze in the Hérault Department.

Following pages: The island of Krk in Croatia. A long history of fire, grazing and wood-cutting, exacerbated by pressure on the vegetation caused by the salt-bearing prevailing winds (the jugo and the bora) which sweep the coasts of Istria has laid bare an extraordinary surface of karst bedrock. The figure of Clara in the photograph, climbing the rocky slope, indicates the scale of this landscape.

pavements criss-crossed by a lattice of regular cracks that harbour a striking and specialized flora arising from influences that are at the same time Mediterranean and alpine, which includes helianthemums, bird's-foot trefoils, thyme and garrigue geraniums growing side by side with gentians and mountain avens. Here too the cutting

Top: Landscape in the Atlas Mountains, showing erosion of soil denuded by cutting down the vegetation and grazing.

Bottom: The fire in the mountains between Poggio d'Oletta and Barbaggio in Corsica in September 2016 was followed by violent autumn rains which opened up erosion gullies that furrow the entire mountain.

of the vegetation and grazing have led to the erosion of the fine layer of post-glacial soil that once covered the rocky slabs; in this region of Ireland, human activities have led to an astonishingly Mediterranean landscape.

The rocky or stony landscapes of garrigues, often very beautiful, are notable for their richness in plants that are able to survive in apparently hostile conditions. Many garrigue plants are perfectly happy in the poorest of eroded soils, where sometimes the only remaining soil is concentrated in narrow cracks in the rocks. In the mountains of the Algarve in southern Portugal, gum cistus (*Cistus ladanifer*) and Spanish lavender (*Lavandula stoechas*) mingle their white and purple flowers on slopes of schist, growing directly on the rock and doing without soil altogether. On the Cycladic island of Sifnos, junipers (*Juniperus phoenicea* subsp. *turbinata*) seed themselves on to limestone blocks, sometimes growing as amazing natural bonsais, squat and twisted, capable of living for decades in a restricting environment. All along the walls of Jerusalem, on the bell towers of Córdoba and on the ochre façades of old monasteries in Sicily, capers cascade down in long dark green sheets, sometimes from a height of tens of metres, far above the ground or any soil, living solely on stones and air. In the spectacular gorge of Topolia in Crete, sages (*Salvia pomifera*), euphorbias (*Euphorbia*

dendroides), ebenus, centaureas (*Centaurea argentea*), the thistle-flowered *Ptilostemon chamaepeuce* and countless bell-flowered *Petromarula pinnata* make a magnificent stone garden, varied and full of flowers, living on the bare rock all the way up the towering vertical sides of the gorge.

• *Oxygen for roots*

For almost 30 years, Clara and I have been studying the potential of plants to acclimatize from the garrigue to the garden. Accustomed to growing in a stony environment, these plants have qualities that make them particularly desirable in gardens – frugality and drought resistance certainly, but also beauty of foliage, enchanting scents and brightly coloured flowers that attract a host of insects beneficial to the gardener. But how should one go about cultivating the ebenus (*Ebenus cretica*) with its magnificent silky leaves that grows in the gorges of Crete? How can one avoid the premature deaths of sages, cistuses and santolinas after only a few years in the garden? What growing conditions does the caper require to produce its delicious edible buds and lovely flowers? Are rosemaries and lavenders able to live as long in the garden as they do in the wild? There is a simple answer to all these questions. In the many gardens around the Mediterranean that we have studied, the main factor influencing the acclimatization of plants from stony garrigue landscapes is the quality of the drainage.

Plants that come from stony garrigues require good drainage because they need the soil to be aerated. Most garrigue plants have roots that explore a large area of the soil: having an extensive root system is one of their main strategies to withstand drought. The roots of a cistus or rosemary may be more than 10m (33ft) long, while those of a kermes oak or Phoenician juniper are able to snake between the stones and into cracks in the rocks for several dozen metres from the plant. In recently eroded soils around old Phoenician junipers one may see exposed roots, as thick as the plant's branches, spreading over the

soil at some distance from the plant. The development of an extensive root system, vital if the plant is to survive in a dry and poor environment, is aided by porous soil; the delicate tips of young roots meet less resistance to their progress if they are able to follow the spaces between the particles of a well-aerated soil. Like all other parts of the plant, roots need oxygen in order to breathe and grow. In compacted clay soils that retain moisture after the autumn and winter rains, the stagnating water replaces the little oxygen available between the fine particles of the soil. By enabling excess water to drain away fast, porous soils by contrast maintain a level of aeration that allows optimal root growth.

The growth of garrigue plants is linked to complex interactions between their roots and a group of soil organisms, including fungi and bacteria. Garrigue species have evolved to survive in poor and stony soil by depending on symbioses with mycorrhizae (from the Greek *myco*, fungus, and *rhiza*, root) that enable them to live in an apparently hostile environment. The symbiosis between roots and mycorrhizal fungi serves multiple purposes. The numerous tiny mycelial filaments of the fungi improve the absorption

On the west coast of Ireland the Burren and the Aran Islands are famous for their limestone pavements scored by a regular network of cracks that are home to a specialized flora, which in this particular damp region of Ireland gives a surprisingly Mediterranean feel to the landscape.

1. *Cistus clusii* growing in the rocky environment of the Sierra Nevada in Andalucia.

2. Many species are able to survive in the apparently hostile conditions of the stony landscapes of the garrigue. Rosemary, common thyme and *Brachypodium retusum* seed into cracks in the limestone cliffs of Saint-Guilhem-le-Désert in the Hérault Department of France.

3. *Phlomis cretica* and *Euphorbia acanthothamnos* in a rocky scree above the Samaria Gorge in Crete.

4. *Ebenus cretica* in autumn, growing on the cliffs of the Topolia Gorge in western Crete.

5. The ruins of Karthea on the island of Kea: adapted to a rocky environment, a cyclamen (*Cyclamen hederifolium*) grows between the stone blocks of a monumental ancient wall.

of water and nutrients by increasing the area over which these exchanges take place in the volume of substrate explored by the roots; every 1m (3¼ft) of root may have up to 1000m (3280ft) of mycelial filaments associated with it. When plants are growing in a rocky milieu, the mycorrhizae also enable the plant to take up nutritive elements inaccessible to the roots alone; they do this by secreting chemical components which attack the bedrock, causing the mineral elements to become soluble so that the plant can absorb them. The rosemaries that cascade down the rocky slopes along the roads at Cap Corse, for example, live in symbiosis with a variety of mycorrhizal fungi which give the plants access to nutrients in this almost entirely mineral environment.

In the case of some species, their tolerance of lime also depends on mycorrhizal symbiosis; where mycorrhizae are absent, the plant shows reduced growth and a yellowing of the leaves. Other garrigue plants exploit symbioses with specialized bacteria that are capable of fixing nitrogen from the air and making it available in the soil at root level.

The elegant *Coronilla juncea* that grows in the calanques of Marseilles and the silky-leaved *Genista linifolia* beside the roads crossing the Sierra Nevada in Andalucia are able to live on the poor and stony slopes thanks to their symbiosis with nitrogen-fixing bacteria in the nodules resembling delicate strings of rosary beads along their roots.

The fungi and bacteria associated with the roots of garrigue plants are aerobic organisms (from the Greek *aer*, air, and *bios*, life) – in other words, they need oxygen in

Top: A garrigue path on the island of Elba. Uncovered by erosion, the powerful roots of a Phoenician juniper snake over the ground at some distance from the plant.

Bottom: Phoenician junipers in the calanques of Marseilles. The Phoenician juniper is able to live in a stony environment thanks to symbioses with mycorrhizal fungi and its roots finding the oxygen they need in fissures in the rock and between stones that make up the scree.

order to live. By limiting a plant's root development and its interaction with micro-organisms, poorly drained soils cause root asphyxiation, leading to the sometimes rapid death of garrigue plants in the garden. In nature, the roots of garrigue plants usually grow in a soil structure composed of larger elements – pebbles, gravel, blocks of stone, in screes or in crevices in the rocks – and in the spaces between these larger elements find the oxygen they need in order to develop, in symbiosis with the micro-organisms on which the plants depend for their growth.

Depending on the nature of the soil and the site, it may or may not be necessary to improve the drainage before starting to plant a garrigue garden. If your garden is situated on a rocky hillside above Nîmes, Granada or Athens, the news is good: thanks to the slope and the stony soil, the drainage is probably already excellent and all you have to do is decompact the soil before planting. The drainage is probably just as good on old terraces supported by dry-stone walls – narrow strips of land won from the garrigue which in the past were planted with rows of vines, almonds or olive trees. Drainage is also excellent on sandy soils which allow water to drain away rapidly after rain, be these on the dunes behind the Hérault coast or outside the Mediterranean zone, for example on the Île de Ré or the Île d'Oléron, where cistuses, arbutuses and *Daphne gnidium* thrive in the wild beside the Atlantic in a distant echo of the garrigue. However, in gardens on clay – especially when it has been compacted by the passage of heavy machinery, as is often the case when a house is built – the success of a future garrigue garden will depend on decompaction of the soil to aerate it before planting and improvement of the drainage to maintain good oxygenation of the soil after rain.

Drainage can be improved in several ways: shaping the land to accentuate natural slopes, adding sand or gravel to raise the level of planting areas, building low dry-stone walls, or creating gravel paths between raised beds which will act as sinks into which water can drain. In our garden,

Teucrium aureum and *Brachypodium retusum* growing on limestone rock. When plants grow in a rocky environment, mycorrhizal fungi enable them to absorb the nutrients that would otherwise be inaccessible to their roots.

which only 30 years ago was an old vineyard on heavy soil, we carried out drainage work before we started planting, and as a result the *Lomelosia* from the cliffs of the Balearic Islands has grown into magnificent cushions; lavenders and rosemaries are ageing well; cistuses are so happy that they have naturalized by self-seeding among the phlomises, sages and euphorbias; capers flower faithfully every year, cascading over dry-stone walls; and the first *Ebenus cretica* that we planted more than 20 years ago in a raised bed still flowers just as spectacularly in springtime.

Depending on the type of soil, ensuring good drainage can involve major work when one is creating a garrigue garden. However, by defining degrees of drainage in different parts of the garden, the gardener can reduce the work involved in the preparation of the soil while at the same time enlarging the plant palette that can be used. Sclerophyllous shrubs with a lignotuber, for instance, are among the garrigue plants that are less demanding when it comes to drainage. In the wild, birds disperse their seeds in all directions, depending on the presence of places where they can perch. In order to make the most of this random scattering of their seeds, often at a distance from the mother plant, an ability to tolerate diverse soil conditions is advantageous to these shrubs; lentisk, phillyrea, buckthorn and viburnums, for example, thrive not only on the poorer soils of rocky hillsides but also on the clay soils of the plains.

Lavenders, sages and santolinas grow in a raised bed to ensure good drainage. Decompaction before planting to aerate the soil and checking the quality of the drainage so that the soil remains properly oxygenated after rain are the two vital stages in the creation of a garrigue garden.

Thus sclerophyllous shrubs which survive fires thanks to their lignotubers can be planted anywhere in the garden, even where the soil is clay, while sub-shrubs native to the stony landscapes of garrigues, such as cistuses, shrubby sages, helichrysums or lavenders, can only acclimatize in areas of the garden where the soil is well-drained. Some plants that are outstanding for the beauty of their foliage or flowers, such as *Sideritis cypria*, *Convolvulus oleifolius*, *Salvia multicaulis* or the caper, need exceptionally good drainage: these plants are only really happy on a stony slope, in a raised rock garden or on top of a dry-stone wall.

• *A new approach to planning*

The distribution of species in the different zones of the garden according to their degree of drainage will to a great extent determine the success of a garrigue garden. How the plants are arranged within each zone is of secondary importance. When planning a traditional garden, gardeners often focus on a precise arrangement of plants, carefully studying colours and flowering periods so that the final picture will be a success from the ornamental point of view. Planning

a garrigue garden requires a different approach. It is not a case of imagining in advance a scene in which each plant will have its own well-defined place, as in an English herbaceous border where the positioning of plants and their successive flowerings are carefully orchestrated; in a garrigue garden the plants will move from year to year, reconfiguring the original planting plan to the point where it may even become unrecognizable after a few years.

Far from becoming progressively less attractive as it diverges from an ideal model, it is as it evolves that the garrigue garden reveals its full interest. By self-seeding freely, the plants find the place that suits them best; the way their positions change simply corresponds to a natural evolution of the garden. Instead of battling against this, it is in the gardener's interest to encourage it; by ensuring the continuity of the plantings, self-seeding reduces maintenance in the long term. In this sense the garrigue garden is a remarkable gardening school, for it invites the gardener to change roles: instead of trying to dominate nature in the garden, he or she can begin to observe the ever-changing nature that is now the daily background to life. The

Top: A garden designed to allow water to run off better: beds raised slightly by the addition of sand and gravel, and paths that help to drain away excess water.

Bottom: Some garrigue plants, such as *Sideritis cypria*, are so beautiful that they merit the provision of conditions that are similar to those of their native habitat – that is, a poor and well-drained soil.

Top: Random placing of plants in a bed: in a garrigue garden plants move around from one year to the next, altering the initial planting plan to such a degree that after a few years it may no longer be recognizable.

Bottom: A garrigue garden provides an opportunity for the gardener to change role: instead of seeking permanent control over nature, he or she may instead focus on observing how the daily environment changes.

gardener discovers the qualities of different plants and the way their beauty changes over the seasons; the multiple sensory experiences they offer from their scents, textures and colours of foliage; the daily magic of the interactions between flowers and the insects that pollinate them; the adaptability of plants to local conditions, and the way they behave, flourishing or not, in different areas of the garden – the progressive creation of a garrigue garden is an apprenticeship that becomes richer and richer.

How to manage plant dynamics and a database to guide selection

A great many garrigue species will self-seed in the garden when they find the conditions that suit them. The gardener can control this ability to self-seed. From the moment of planning the garden, the careful choice of species enables him or her to guide the future evolution of the garrigue garden: some pioneer plants spring up all over the garden, while other species remain obediently in their allotted place. Even within a single genus, species may behave very differently: the Cretan cistus (*Cistus creticus*), the downy cistus (*C. albidus*) and their natural hybrid (*C.* × *canescens*) self-seed in stony soil, whereas other hybrids such as *C.* × *purpureus* and *C.* × *pulverulentus*, often grown in gardens for their ornamental qualities, are sterile and do not self-seed.

The way the soil is prepared and the choice of mulch will also define which species will be able to self-seed. Ruderal herbaceous plants such as mallow and sticky fleabane self-seed easily in heavy soils, while Cupid's dart (*Catananche caerulea*), love-in-a-mist (*Nigella damascena*) and hollow-stemmed asphodel (*Asphodelus fistulosus*) prefer poor and stony soils. The application of an inorganic mulch strongly influences the way the garden will evolve; it limits the germination of weeds while favouring the self-seeding of heliophile garrigue sub-shrubs such as lavenders and santolinas. The use of allelopathic plants, which give off chemical compounds that inhibit the germination of other competing species (see p.175), also reduces the appearance of self-sown seedlings in certain areas of the garden, for example among groundcover plantings where one wants to minimize weeding.

When planning a garrigue garden, the gardener may find himself or herself asking new questions about the way the chosen plants will evolve. How rapidly do they grow and how long do they live? Will they spread to other spots or will they remain where they were originally planted? What is their ability to self-seed depending on the soil type? From our observations in the wild and in numerous gardens we have put together this information on growing garrigue plants in a database, available on the internet (see p.276). This database allows the gardener to choose species according to several criteria: their drainage requirements, their adaptation to local conditions (hardiness, drought tolerance, lime tolerance, position), their life span and rate of growth, their self-seeding potential depending on the nature of the soil and their various functions in the garden (honey plants, plants that attract beneficial insects or birds, allelopathic plants). The search engine allows the data on a plant's behaviour to be cross-referenced with other data concerning the functional or aesthetic aspects of the plant: its use (groundcover, mixed hedge, gravel garden), its shape (ball or cushion shapes, vertical silhouettes, carpeting plants, climbing or cascading plants), its foliage (deciduous or evergreen, aromatic, grey, green or silver) and its flowering period and flower colour. Applying these successive filters makes it much easier to choose species suitable for the different areas of the garden.

This database was conceived in order to widen the horizons of gardeners during the planning of a garden: by drawing inspiration from their long history of adaptation to fire, to human interventions or to erosion, gardeners can select garrigue plants not only for their contribution to a variety of ornamental scenes but also as the elements of an ecosystem with an evolution which the gardener may use to guide different zones of the garden.

EVERGREEN FOLIAGE: RICH IN CONTRAST, COLOUR AND TEXTURE

Page 92: The garrigue
garden in winter is
enriched by a diversity
of foliage colours
and textures.

Right: Evergreen
foliage provides a
structural framework
to ensure the garrigue
garden is beautiful not
only when plants
are in flower but also
during periods
when there are
fewer flowers.

The garrigue garden has one special aesthetic quality: it remains visually attractive all through the year independently of its plants' flowering seasons. The garrigue includes many evergreen plants, which have a vegetation cycle adapted to the demands of the mediterranean climate. To get through the difficult summer season, the evergreen plants become dormant when the weather is hot and dry. They limit water loss during this period by reducing the gaseous exchanges connected with photosynthesis, which causes a pronounced slowing of growth in summer. On the other hand, they retain their leaves in winter: after their summer rest they resume growth in autumn and continue to grow through winter and spring, creating new interest for the gardener. We are very far from the classic idea of a garden as a place where a floral display counts above all else; thanks to its diversity of foliage, the garrigue garden is beautiful not only when plants are in flower but also during other times of year.

All around the Mediterranean, evergreen plants create remarkable landscapes, striking for their diversity of colours and textures. On the east coast of Cap Corse, between Centuri and Nonza, vast undulating expanses shaped by the wind form a long green fringe along the coast in which the luminous green of tree heathers mingles with the glossy dark green of lentisks, which take on bronze tints in winter. At the Bay of Naples, as one goes down towards Punta Campanella at the tip of the Sorrento peninsula, the light green balls of the winter foliage of tree euphorbias (*Euphorbia dendroides*) progressively give way to a landscape that is at first silver and then pure white, consisting of Jove's beard (*Anthyllis barba-jovis*),

silky-leaved *Convolvulus cneorum* and *Centaurea cineraria* with its delicate ash-grey leaves. On the Cycladic island of Kea, the path that leads to the tiny chapel of Agios Nikolaos abutting the lighthouse at Korissia is bordered by the curious interwoven masses of thorny burnet (*Sarcopoterium spinosum*), centaureas (*Centaurea spinosa*) and broom (*Genista acanthoclada*), with spiny stems ranging through all shades of grey-green, silver-grey and golden-grey in summer. In the High Atlas, as you cross the Tizi N'Test pass before embarking on the long descent to the Sous plain which lies below, foundering under the heat, light and dust, you come upon a striking landscape in which all the plants have glaucous foliage: the local variety of dwarf palm (*Chamaerops humilis* var. *argentea*) spreads in large metallic-blue clumps, among which are slender tufts of *Linaria ventricosa*

Teucrium marum in our garden. Far from being constraints, lack of water, wind and poor soil can be real assets in a garrigue garden: they accentuate the beauty of the foliage and allow plants to grow in natural conditions where they can express their resistance strategies.

with grey-blue leaves and squat bushes of *Adenocarpus anagyrifolius* with magnificent silvery-blue foliage.

The many variations in the appearance of the foliage of evergreen garrigue plants correspond to the multiple strategies they use to adapt to tough living conditions. The plants must withstand not only summer drought but also strong winds, stony soil, cold at higher altitudes, salt in coastal regions and the continual pressure exerted by herbivores. Thus in a garrigue garden, far from being handicaps that limit the gardener's possibilities, lack of water, wind, burning sun and poor soil become real assets: by providing plants with a situation that is natural to them in which they can express all the elegance of their survival strategies, these factors simply accentuate the beauty of their foliage. Being perfectly adapted to difficult growing conditions, evergreen plants are an inexhaustible source of material for the creation of a lasting framework, rich and complex, with a particular beauty that is the mark of a garrigue garden. The colour palette of this framework is exceptional, ranging from the darkest of greens to pure white, via all the different shades of light green, gold, blue and silver-grey. When one is planning a garrigue garden foliage colours are of primary importance. Flower colour is of course important too, but it takes second place when one is choosing plants and deciding on their positions in the garden; the flowering seasons of garrigue plants are often brief, while the structural foliage is present throughout the year, forming the very heart of the garden.

• *Glossy leaves, the dark green of garrigues*

Sclerophyllous plants, so numerous in garrigues, have thick and leathery dark green evergreen leaves. The epidermis of their leaves is covered with a glossy cuticle which reduces water loss by evaporation and gives the plants their fine dark colour. These plants are very useful in gardens as their foliage remains attractive during the difficult summer months; in a garrigue garden they help to form an ornamental summer framework whose visual beauty is independent of drought or heatwaves. Myrtle (*Myrtus communis*), for example, used since antiquity in Mediterranean gardens, is a sclerophyllous shrub of interest for its countless small white flowers in June, followed by decorative berries in autumn; however, its greatest interest lies in its dark green evergreen leaves, deliciously aromatic, which are as attractive in summer as they are in winter. Narrow-leaved mock privet (*Phillyrea angustifolia*) and lentisk (*Pistacia lentiscus*), shrubs that are emblematic of garrigues in the South of France, are also unbeatable in a garrigue garden: undemanding as regards cultivation conditions, these are robust shrubs with foliage that remains a handsome glossy green in winter as in summer, even in periods of exceptional heat and drought.

Sclerophyllous plants are found all round the Mediterranean and belong to various biological types. These include sub-shrubs such as the Gibraltar shrubby hare's-ear (*Bupleurum gibraltaricum*), which grows on the steep slopes of the Sierra Nevada in southern Spain, large shrubs such as buckthorn (*Rhamnus alaternus*), common in the garrigues of the South of France, or trees such as the carob (*Ceratonia siliqua*), the Palestine oak (*Quercus coccifera* subsp. *calliprinos*) and the magnificent Cyprus arbutus (*Arbutus andrachne*), seen throughout the Eastern Mediterranean.

Left: Myrtle (*Myrtus communis*) is of interest for its white flowers in June, but also for its dark green, deliciously aromatic, evergreen foliage which is equally attractive in summer and winter.

Right: The common arbutus (*Arbutus unedo*): a contrast in winter between its red fruits, white flowers and dark green evergreen foliage.

Bottom: Sclerophyllous plants form a decorative structure in summer with visual qualities that remain through drought and heatwaves. The Park of Saleccia, near L'Île-Rousse in Corsica.

The mountains of Mallorca are home to a diversity of sclerophyllous species with especially beautiful foliage. Balearic St John's wort (*Hypericum balearicum*), which forms compact balls on rocks above the sea, has small, leathery, glossy dark green leaves with a wavy margin. Balearic buckthorn (*Rhamnus ludovici-salvatoris*) has leaves so dark that they seem almost black when the plant stands out in summer amid the supple golden expanses of *Ampelodesmos mauritanicus* between the bare peaks of Puig Major and Puig Roig. The rounded clumps of *Cneorum tricoccon*, with thick, elongated leaves, add rhythm at the foot of the dry-stone walls bordering the cobbled paths that make the landscapes of the Serra de Tramuntana so striking. Balearic boxwood (*Buxus balearica*), with large rounded leaves of a handsome shining green, forms twisted bushes that cling to the ochre-

Left: Sclerophyllous oaks (*Quercus coccifera* subsp. *calliprinos*) grow with olives in the Aradena Gorge in Crete.

Right: The dark shapes of carobs in the landscape behind the dunes on the Gramvoussa peninsula in Crete.

coloured limestone of the cliffs towering above the spectacular gorge of the Torrent de Pareis. Shown off to their best in the landscape design of the botanic garden in the Mallorcan town of Sóller, all these plants deserve to be used more often to create structure in Mediterranean gardens; in spite of their ornamental value, the diverse range of sclerophyllous plants remains largely under-used by Mediterranean gardeners.

• *Reflection and the play of light on leaves*

Sclerophyllous plants create a striking play of light in the garden for, according to the time of day and their position in the garden, they may appear as dark or luminous masses. The leaf tissue absorbs only part of the light energy in the spectrum of visible light and the surplus light energy not needed for photosynthesis is reflected in a scattered manner by the photosynthetic pigments which give leaves their green colour. However, in very bright sunlight sclerophyllous plants take their strategy of reflecting light one step further. As well as the scattered reflection of green light common to all chlorophyll-producing plants, their shiny cuticles produce specular reflection (from the Latin *speculum*, mirror) which substantially reduces the solar radiation reaching the tissues beneath the epidermis. If the solar radiation is too strong, specular reflection prevents an excessive rise in the temperature of the tissues situated beneath the cuticle. Unlike the scattered reflection of green light, specular reflection gives the colour white, for it reflects back all light rays in the visible spectrum: dark green leaves suddenly appear a dazzling white when they catch the rays of the sun.

The cuticles of sclerophyllous plants thus serve several functions: they reduce evaporation in times of hydric stress and reflect back excessive light energy which could damage the cells in plants with more tender leaves. Thanks to these cumulative strategies, most sclerophyllous plants have a remarkable ability to adapt to tough climate conditions. In the wild, for example, the laurel-leaved cistus, recognizable from its thick, leathery, dark green leaves sometimes covered with a silvery bloom, grows in very varied environments, from the snow-covered slopes of the Western Taurus in Turkey to the windswept passes of the High Atlas in Morocco via the burning plains of La Mancha near Albacete in southern Spain.

To complete the visual interest of the play of reflections and light, sclerophyllous plants often show two colours on their leaves; the upper surface is dark green, while the underside has a lighter colour. This difference in colour, more or less marked according to the species, is linked to the position of the stomata, those minute openings through which oxygen and carbon dioxide are exchanged during photosynthesis: the upper surfaces of the leaves are covered with a thick, impermeable varnish, and the stomata are located on their undersides where they are protected from the sun's rays, thus reducing water loss during gas exchange. The leaves of the oleander, for instance, have a glossy dark green upper surface. Their undersides are lighter in colour, being covered with numerous tiny white dots, which you can see if you examine a leaf under a simple naturalist's magnifying glass. These white dots are the stomatic crypts, deep cavities whose narrow openings are fringed with a crown of white hairs which act as a filter

Left: *Viburnum tinus* used as a structural plant in a garden.

Right: Dazzling reflections caused by specular reflection are apparent on viburnum leaves when the plant is seen against the light.

Bottom: *Cistus laurifolius* subsp. *atlanticus* in the High Atlas in Morocco. The cuticles on the leaves of sclerophyllous plants serve several functions: they limit evaporation during periods of hydric stress and reflect back excess light energy that could damage the leaf cells.

to maintain humidity around the stomata situated at the bottom of the crypts, where they are protected from wind and extreme heat.

In south-west Crete, on the little road that winds through the spectacular gorge of Kotsifos, you may see a sclerophyllous plant with particularly fine foliage, *Staehelina petiolata*. If you lean over the safety barrier to look at the plants below the road, you will see the staehelinas growing on the cliffside in regular balls of dark green leathery leaves, each edged with a fine silver border. But if you turn round and look upwards at the cliff towering above the road the staehelinas appear to be silvery, because from below you see only the light colour of the undersides of their leaves. The upper surfaces of the leaves are glossy and dark as a result of the protective cuticle which enables the plant to survive in the tough conditions of these baking cliffs, while the silvery colour of the undersides is due to a

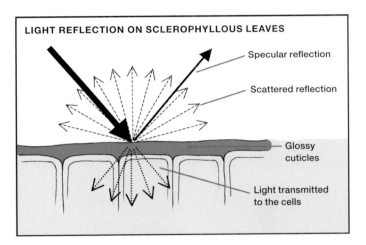

LIGHT REFLECTION ON SCLEROPHYLLOUS LEAVES

Specular reflection

Scattered reflection

Glossy cuticles

Light transmitted to the cells

Top: Apart from the scattered reflection of green light common to all chlorophyll-producing plants, the glossy cuticles of sclerophyllous plants create specular reflection, which reduces the solar radiation reaching the tissues beneath the epidermis.

Bottom: Oleanders in Crete, growing in the bed of a watercourse that is dry for many months of the year.

thick felting of hairs protecting the stomata. This juxtaposition of two colours on the leaves of many sclerophyllous plants, curious and decorative, is part of what makes garrigue gardens so striking. It is best seen when there is a strong wind, frequent in the Mediterranean. In the South of France, when the mistral or the tramontane blows, the lighter undersides of the leaves of arbutuses and viburnums are briefly revealed, the leaves of the holm oak show

their pale grey undersides, and olive trees are transformed into magnificent, ever-moving silver masses, ceaselessly shaken by the violent gusts.

• Reducing leaf surface in summer: a strategy of plants with light green foliage

Garrigue plants with light green foliage have to call on different strategies from those of sclerophyllous plants to resist drought, for their soft leaves, lacking a shiny cuticle, are less well protected from water loss. Many such plants minimize water loss during the dry period by simply reducing their total leaf surface. In Sicily, in the Nebrodi Mountains which form a barrier in the north of the island, the rocky pastures are colonized by tight clumps of

giant fennel (*Ferula communis*). Their unpalatable young shoots, carefully avoided by sheep, form an even landscape of beautiful soft green at the end of winter, but after flowering the rich, supple, finely cut leaves disappear, leaving only the tall stems of their dry inflorescences to mark the landscape all summer long, the plants surviving thanks to their thick taproots anchored between the stones, which act as a reserve through the dry season.

The tree medick (*Medicago arborea*), common in the coastal garrigues of Greece and southern Turkey, regulates its total leaf surface according to the amount of water available, maintaining some of its leaves in cooler summers and becoming almost entirely deciduous in periods of extreme dryness. The handsome luminous green bushes

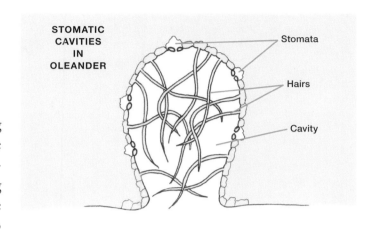

STOMATIC CAVITIES IN OLEANDER

Stomata

Hairs

Cavity

of stinking bean trefoil (*Anagyris foetida*) that grow along the paths on the slopes of Mount Parnassus around the ruins of Delphi are transformed in summer into unassuming clumps of dry wood. At the end of summer, on the coasts of Italy the branching shapes of bare tree euphorbias standing out amid golden expanses of tall grasses (*Hyparrhenia hirta* and *Ampelodesmos mauritanicus*) bear witness to the sudden change in the colours of

Top: In oleanders the stomata are hidden in cavities, or stomatic crypts, on the undersides of the leaves. The hairs surrounding these crypts reduce water loss during the gas exchanges involved in photosynthesis.

Bottom: The Nebrodi Mountains in Sicily – a light green landscape dominated by the new shoots of *Ferula communis* in late winter.

Following pages: A day when the meltemi is blowing on the Cycladic island of Naxos. Tossed by the wind, thousands of olive trees show the silvery undersides of their leaves.

Tree euphorbias in late summer on the Bay of Naples emerge from yellow expanses of *Hyparrhenia hirta* and *Ampelodesmos mauritanicus*.

the landscape that will occur as soon as the first autumn rains arrive. In a garrigue garden, plants with light green foliage which lose all or some of their leaves in summer can be used as a counterpoint to dark green sclerophyllous plants, the movement of their changing appearance over the seasons completing and enriching the permanent structure of the garden.

• *The wax crystals behind blue, blue-green and glaucous foliage*

Some garrigue plants that are particularly attractive in the garden have foliage in an amazing range of shades of blue, grey-blue and silvery-blue. *Linum arboreum*, growing in crevices in the rocks of the White Mountains in Crete, has leaves of a bluish colour that sets off the luminous yellow of its delicate bell-shaped flowers. On the Aeolian Islands to the north of Sicily, *Dianthus rupicola* is a shrubby pink with a striking bushy habit, as notable for its handsome bluish leaves as for its long flowering season. In Catalonia, the vegetation of Cap Creus, sculpted by wind and salt, is given rhythm by the alternating dark green of Phoenician junipers (*Juniperus phoenicea* subsp. *turbinata*) and grey-blue of prickly junipers (*J. oxycedrus*). During summer, nonchalant cows lie on the carpet of seagrass fruits on the beaches of Cap Corse between Barcaggio and Centuri, in a curiously blue landscape dominated by sea holly (*Eryngium maritimum*), which has stems and prickly leaves of a magnificent bright mauve-blue; and on the limestone slopes of the M'Goun massif in the north of the Atlas Mountains, the grey-blue stems of delicate bushes of *Polygala balansae* stand out among cactiform euphorbias

Left: The violet-blue stems and leaves of the sea holly, *Eryngium maritimum*.

Right: The delicate bushes of *Polygala balansae*, with blue-grey stems, emerge from tight masses of silvery-blue cactiform euphorbias (*Euphorbia resinifera*) on the northern slopes of the M'Goun Atlas.

Bottom: Cows and sea holly on a beach at Cap Corse.

Top: A silver-blue layer of euphorbias (*Euphorbia officinarum* subsp. *echinus*) and the dark shapes of argan trees (*Argania spinosa*) in the arid environment of the Anti-Atlas.

Bottom: The spiral arrangement of the silver-blue leaves of *Euphorbia rigida* allows maximum light absorption.

• *Architectural structure and light*

Plants with blue foliage often have an architectural shape which gives them added visual interest in the garden. The leaves of *Euphorbia rigida*, for example, are arranged in a regular spiral around the stems. The thick, impermeable layer of silvery-blue wax that covers the leaves reflects light powerfully, reducing photosynthesis and enabling the plant to withstand long periods of drought without problem in the mountain garrigues of the Eastern Mediterranean Basin, for instance in Sicily, in the Peloponnese and in the Western Taurus in Turkey.

In compensation for this reduced photosynthesis, its spiral architecture allows the euphorbia to increase its overall light-absorbing surface. The arrangement of its leaves enables it to catch the light with maximum efficiency: the position of each leaf is staggered in relation to the leaf above it, the angle of divergence between two successive leaves in the spiral, called 'the Fibonacci Spiral', being large enough to prevent each leaf casting shadow on the next. Thanks to the high coefficient of reflection of the wax covering the epidermis of the leaves, the spiral architecture also increases light absorption by the leaves' undersides: the light reflected by the wax crystals on the surface of a leaf shines in part on the underside of an adjacent leaf. Such successive rebounding of the light rays means that all

(*Euphorbia resinifera*) with monumental cushions that form magnificent waves of silvery-blue.

Plants with blue foliage have resistance strategies similar to those of sclerophyllous plants. Their thick leaves exude an impermeable wax, the purpose of which is to reduce water loss and to reflect light. Thanks to the variety of microcrystals of which this wax is made, which create an impressive range of different shades of blue, the specular reflection coefficient of blue leaves is particularly high. The wax is able to reflect a wider spectrum of excess light energy than the cuticles of sclerophyllous plants, particularly in the ultraviolet range of wavelengths, which gives them an excellent ability to adapt to intense solar radiation and the toughest of living conditions. In southern Morocco, where garrigues are evolving into arid lands baked by the sun, representing the last stage before desertification, the remarkable beauty of the landscapes of the Anti-Atlas derives from the contrast between the silvery-blue layer of steppe covered with *Euphorbia officinarum* subsp. *echinus* and the dark silhouettes of argan trees (*Argania spinosa*), capable of surviving even the most severe droughts.

parts of the silver-blue leaves benefit from direct or indirect light, the colour and striking architecture of the plant increasing the total surface available for photosynthesis by a remarkable optimization of light absorption.

• *Down, wool and silk: the beauty of grey foliage*

To complete the palette of colours and textures available to the gardener, many garrigue plants have leaves that are downy or silky, coloured grey-green, silver-grey, golden-grey or sometimes almost white. On the edge of the Omalos plateau above the Samaria Gorge in Crete, an oregano with tiny silver leaves (*Origanum microphyllum*) grows alongside the soft grey leaves of *Stachys cretica*, the silky white cushions of mountain tea (*Sideritis syriaca*) and the rounded hummocks of a phlomis with thick and woolly golden-grey leaves (*Phlomis cretica*). Below the Larzac plateau, all along the stony paths that cross the garrigues of Saint-Guilhem-le-Désert, *Teucrium aureum* forms delicate, smooth-looking silver carpets, which are part of the structure of a natural rock garden where the leaves blend into the environment; tucked among the grey, white or orange-tinged limestone rocks are the grey-leaved *Phlomis lychnitis*, *Helichrysum stoechas* with narrow silver leaves, the cottony grey-green leaves of *Staehelina dubia*, the silky silvery balls of *Mercurialis tomentosa* and grey clumps of aspic lavender (*Lavandula latifolia*).

In the Sierra Nevada, in Andalucia, the aspic lavender is replaced by woolly lavender (*Lavandula lanata*), with leaves covered in a fine ash-grey fleece. Woolly lavender grows at higher altitudes in poor and stony soil, together with *Cistus albidus*, which has leaves covered in a fine grey down, *C. atriplicifolius* with silvery leaves, and purple phlomis (*Phlomis purpurea*), the grey-green winter foliage of which becomes almost white during long periods of summer drought. Having come down from the Sierra Nevada passes to the sea, one finds the common rosemary of the garrigues (*Rosmarinus officinalis*) growing on the few rocky promontories that have escaped urbanization,

To limit water loss, the leaves of *Rosmarinus × mendizabalii* are covered in a thick silver down.

in company with a much rarer species, the silver-leaved *Rosmarinus tomentosus*. These two species interbreed freely, giving rise to a striking hybrid, Mendizabal's rosemary (*Rosmarinus × mendizabalii*), with vigorous silvery clumps that cling to the rocks above the sea.

Grey-leaved plants are one of the greatest assets in a garrigue garden. In our own plot we have been progressively expanding the part that we call our 'gravel garden', which has over the years become one of our favourite experimental areas. This unusual-looking part of the garden is home to diverse plant types with a variety of shapes and volumes, giving what a cinematographer might call a dissolving shot of grey tones and textures that are mostly soft, downy, silky, cottony or woolly. The finely cut silver foliage of large artemisias (*Artemisia thuscula* and *A. arborescens*) gives structure to the space with its imposing mass. The tight balls of Albaida broom (*Anthyllis cytisoides*), tree germanders (*Teucrium fruticans*) and cistuses (*Cistus halimifolius*, *C. × tardiflorens*) complete the silver-grey structure of taller plants. An intermediate layer consists of ballotas with grey-green or silver-grey leaves (*Ballota hirsuta*, *B. pseudodictamnus* and *B. acetabulosa*), silvery lavenders (*Lavandula × chaytorae*), grey-green salvias (*Salvia fruticosa*, *S. pomifera*) and many *Phlomis* species with grey foliage that often takes on russet or golden tints at the hottest time of summer.

Top: A garden
dominated by grey
and silver foliage.

Bottom: The golden
form of the
Portuguese germander
(*Teucrium
lusitanicum* subsp.
aureiforme) makes a
silvery scene, like a
miniature garden,
beside the road in the
Sierra Nevada, in
southern Spain.

The layer of plants that form low cushions or carpeting groundcover includes various garrigue germanders (*Teucrium marum*, *T. cossonii*, *T. luteum*), lavender-leaved salvias (*Salvia lavandulifolia* subsp. *oxyodon* and *S. lavandulifolia* subsp. *vellerea*), artemisias (*Artemisia pedemontana*, *A. herba-alba*) and thymes (*Thymus leucotrichus* subsp. *neiceffii*, *T. munbyanus* subsp. *ciliatus*). To give emphasis to the dynamic evolution of this part of the garden we have planted a selection of pioneer species which self-seed profusely, creating each year new sequences of grey which can reveal stunning chance associations: *Dorycnium pentaphyllum* and *D. hirsutum* seed themselves into irregular borders along the paths through the planting, downy-leaved salvias (*Salvia sclarea* and *S. candelabrum*) colonize the spaces between the structural plants, and horned poppies (*Glaucium flavum*), luminous and hairy,

self-seed as soon as the earth is disturbed, forming a silver screen around new plantings. The overall scene is set off by a few touches of dark foliage (*Pistacia lentiscus*, *Cneorum*

tricoccon, Ononis speciosa, Erica manipuliflora, Silene fruticosa) and by the glaucous foliage of euphorbias (*Euphorbia rigida, E. nicaeensis, E. myrsinites, E. characias* subsp. *wulfenii*), which self-seed with an unequalled generosity and thus contribute to the dynamic of pioneer plants.

• A coat of insulating hairs

The exceptional diversity of grey-leaved garrigue plants comes from their range of sizes, shapes and colours as well as from the density of the hairs that cover the green tissue of their leaves. The leaves of downy plants are in fact green, although one might not often guess it: the silver of Dubrovnik centaury (*Centaurea ragusina*), for example, is nothing more than a thick network of minute interwoven hairs that cover the green epidermis of each leaf, the green pigments being necessary for photosynthesis. These hairs, which botanists call trichomes (see the box on p.112), have multiple functions in helping the plant to withstand the tough conditions of the garrigue: they play a part in reducing the stresses of dryness, heat, wind and intense sunlight. In some cases they also have a particular role which gives the plant a decisive competitive advantage, allowing a limited number of hairy species living in a hostile environment to become dominant over an entire landscape, contributing to the power and special beauty of some of the most remarkable garrigue scenes.

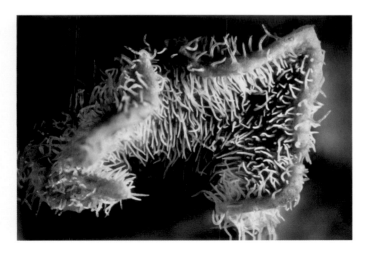

Like the shiny cuticles on the leaves of sclerophyllous plants or the waxy microcrystals on glaucous leaves, the hairs on silky or downy leaves reflect back the solar rays, including those in the ultraviolet and infrared spectra, protecting the leaf from an excessive rise in temperature. The leaves of Jove's beard (*Anthyllis barba-jovis*), for example, are entirely covered in fine silver silky hairs that reflect back the light; it is the varying orientation of these silky hairs that causes the lovely shimmering appearance that makes the plant so elegant. *Inula verbascifolia*, precariously growing in the limestone screes above the Dalmatian coast, forms pure white clumps that stand out against the dark background of the sea, its leaves being covered in white hairs which reflect the light perfectly.

All around the Mediterranean there are numerous grey-leaved plants in regions where the sunlight is particularly intense, such as in transitional zones between the garrigue and high mountains or desert. The miniature lacy leaves of *Achillea umbellata* are covered in a fine network of minute hairs, invisible to the naked eye, which protect the plant from the extreme solar radiation found at high altitudes, helping it to survive on the bare peaks of the mountains of the Peloponnese. The leaves and stems of *Teucrium capitatum*, which grows in the austere landscape of the rocky slopes near the Kerdous pass in the Anti-Atlas, are covered in minute cottony white hairs that shelter the plant from the brutal rays of the sun. Along the road in Israel that leads down to the Dead Sea from the desert plateaux of the Negev, the sea orach (*Atriplex halimus*) is one of the last plants capable of surviving in this burning hot and salty environment, for the curious vesicular hairs that cover the leaves and give them their silvery-grey colour reflect back a significant amount of light energy – up to 60 per cent of solar radiation.

The trichomes sometimes take surprising forms which serve further protective functions. In some grey-leaved species, such as dittany of Crete (*Origanum dictamnus*), clary sage (*Salvia sclarea*) and *Stachys byzantina*, the hairs

Detail of a leaf of the horned poppy (*Glaucium flavum*): beneath their coat of hairs, grey and silver leaves are in fact green (magnification × 30).

The Peljesac peninsula in Croatia, with the island of Korčula in the background. The white-leaved *Inula verbascifolia*, perched in the limestone screes above the coast, has its leaves covered in a tight down that reflects the light.

are so long that they lie horizontally and form a dense network like a false ceiling above the leaves, trapping a narrow layer of air between the epidermis and the veil of protecting hairs. This stationary layer of air, protected from the climate conditions outside, maintains a cooler temperature and a greater degree of humidity in summer, sheltered from heat and wind. This humid environment immediately above the epidermis enables the plant to lose the least amount of water possible through the stomata during the gas exchange involved in photosynthesis. The layer of air can also serve in winter to insulate the plant from cold on mountains. Many species that grow on Mediterranean mountains adopt this dual strategy to withstand both drought and cold by enveloping themselves in a dense blanket of hairs that acts as an insulating filter. Hairy phlomis (*Phlomis crinita* subsp. *malacitana*), for example, which grows high up in the mountains of southern Spain, forms clumps of silver leaves clad in a thick coat of hairs that allows the plant to withstand both the heat of summer and the cold of winter.

Top left: The leaves of *Phlomis crinita* subsp. *malacitana*, which grows high in the mountains of southern Spain, are covered in a thick coat of hairs which enable it to withstand both heat in summer and cold in winter.

Top right: Detail of the silky hairs on the inflorescences of *Ebenus cretica* (magnification × 40).

Bottom: Contrasting shapes, textures and colours: *Lavandula dentata* var. *candicans*, olives and cypresses.

1. A fine, supple hair with a downy texture (*Origanum syriacum*).
2. Long hairs lying flat with a silky texture (*Stachys cretica*).
3. Stiff, pointed hair with a rough texture (*Onosma albo-roseum*).
4. Candelabrum-shaped hair with a velvety texture (*Phlomis russeliana*).
5. Star-shaped hair with a felty texture (*Ballota acetabulosa*).
6. Intertwined hairs with a woolly texture (*Dorycnium hirsutum*).
7. Sessile glandular hair: aromatic leaves (*Salvia pomifera*).
8. Glandular hair with a head: aromatic and sticky sepals (*Salvia candelabrum*).
9. Hair equipped with curved barbs: prickly leaves (*Pieris hieracioides*).
10. Tree-shaped hair with a woolly texture (*Phlomis nissolii*).
11. Fine and regular hairs with a feathery texture (*Ebenus cretica*).
12. Parasol-shaped hair (*Olea europaea*; 12a cross section; 12b seen from above).

Hairs, glands and trichomes: the carnival of the garrigue

Trichomes (from the Greek *trichoma*, 'growth of hair') clothe grey-leaved plants in a fascinating top layer. Under a stereoscopic microscope elaborate miniature landscapes can be seen, often beautiful, curious or extravagant, which reveal the plants' incredible ability to adapt to the tough conditions of the garrigue. The simplest trichomes, which resemble mammalian hairs, can be stiff and short, as on the leaves of the horned poppy (*Glaucium flavum*), long, fine and silky, as on the backs of the sepals of the Montpellier cistus (*Cistus monspeliensis*), or gathered together in regular little clumps, for example on the leaf stalks of *Salvia interrupta*. These hairs, usually unicellular, may be hollow in order to increase their ability to reflect light: in this case they give a pure white colour, as on the leaves of *Inula candida*. Sometimes they have more complex forms, with the hairs branching to make a star shape (*Ballota pseudodictamnus*), balancing on one leg, as it were, to take on a tree-like form (the dendroid trichomes of *Phlomis* species) or branching into a Y shape or a multi- branched candelabrum (the tiered trichomes of mulleins). Some star-shaped hairs have a dominant central branch that is slender, aggressive and erect, like a sharpened fencing foil (prickly-leaved plants of the bugloss family). Other star-shaped hairs take interwoven forms, creating a tangle of spikes and points whose purpose is to block the way to insect pests or to keep them at a sufficient distance from the epidermis so that their rostra cannot reach it (*Phlomis nissolii*). Other trichomes resemble parasols or shields borne on a short leg, as for example the scale-like hairs on the backs of olive leaves.

A very large group of trichomes comprise glandular hairs, veritable factories producing chemical compounds that help the plant survive in particular conditions. These glandular hairs may resemble a small spherical head perched on a long leg (*Salvia candelabrum*), a large mushroom-shaped head on a squat leg (*Thymus vulgaris*), or a shiny bead which appears to be placed on the leaf surface itself, often protected by an upper layer of fine hairs (*Lavandula angustifolia*).

HAIR TYPES OF GARRIGUE PLANTS

• *Hairs that resemble trees*

In order to better trap an insulating layer of air in immediate proximity to the leaf's epidermis, trichomes sometimes have curious tree-like forms. The dendroid hairs (from the Greek *dendron*, tree) that cover the magnificent golden leaves of the Atlas mulleins (*Verbascum* sp.) are shaped like little branching trees, with the branches arranged in several tiers, touching one another, to create thick 'foliage' which lets air in gently while at the same time protecting the leaf from the heat outside and from desiccating wind. The dendroid hairs of *Phlomis russeliana*, which grows in the mountain ranges of northern Turkey, are of a different shape, with spreading branches above a central peduncle, like diminutive versions of the star at the tip of a fairy's wand. The upright branches of these tree-like hairs, which give the leaves a pleasant velvety texture but are invisible to the naked eye, are of an ideal size to capture the tiniest droplets of morning dew, having a diameter in the order of a hundred microns. In this way the leaf surface is doubly protected, by the tight network of tree-like hairs and by the amazing density of the microdroplets that form a moist roof suspended above the epidermis in the hairs' branches. When the day starts to heat up, the fall in temperature brought about by the evaporation of the dew clinging to the hairs helps for a little while to maintain conditions more favourable to

photosynthesis, the damp hairs thus acting temporarily as a miniature air-conditioning system. In periods of extreme hydric stress, some downy garrigue plants such as Jerusalem sage (*Phlomis fruticosa*) even manage to 'drink' a small amount of water through their leaves; in this way the occult precipitation (unmeasured moisture) from summer dew and the first autumn fogs, trapped by the hairs, can help grey-leaved plants get through the most severe periods of drought as they wait for the rainy season to arrive.

• *Daggers, bayonets and barbed wire*

Although most silky or velvety trichomes give leaves such a soft appearance that one wants to touch them, some of the hairs on garrigue plants end in points or hooks which give the leaves a rough or prickly texture. These sharp trichomes serve a very specific purpose: defending the plant against attack from insects, molluscs and herbivores. The leaves, stems and calyces of *Lithodora hispidula*, which grows near the little port of Sougia in south-west Crete, are covered in minute rigid and pointed hairs resembling short daggers, arranged like a portcullis to block the advance of aphids and other sap-sucking insects. The grey-green leaves of *Onosma albo-roseum* bristle with long, stiff hairs, as sharp as bayonets, ready to impale the soft bodies of any caterpillars, slugs or snails that venture

Left: The rosette of a mullein (*Verbascum* sp.) near the Tizi N'Test pass in the Atlas; the thick fleece of golden hairs filters the outside air, enabling the plant to live in very tough conditions.

Right: Droplets of dew trapped by the tree-shaped hairs of *Phlomis russeliana*. Some downy garrigue species manage to 'drink' a small amount of water through their leaves, so that the nocturnal dew of late summer, trapped by the hairs, can enable grey-leaved plants to get through even the severest droughts as they await the rainy season (magnification x 40).

Left: Like many plants of the bugloss family, *Onosma albo-roseum* is resistant to attack by insects because its leaves, stems and inflorescences are protected by sharp hairs.

Right: The bristling bayonets on the leaves of *Onosma albo-roseum*, ready to impale caterpillars or snails, are surrounded by a crown of small secondary hairs, just as pointed, which block the way to insects and mites (magnification × 40).

Opposite top: Gum cistuses (*Cistus ladanifer*) and butterfly or Spanish lavender (*Lavandula stoechas*) create a bouquet of scents on the schist hills of southern Portugal.

Opposite bottom: The lavender-leaved sage (*Salvia lavandulifolia* subsp. *vellerea*) makes an aromatic carpet in the Sierra Nevada in Andalucia. The essential oils given off by the leaves protect the plant from herbivores and inhibit the germination of competing species.

on to the plant. To discourage the microscopic insects and mites which might try to slip between the close ranks of bayonets, the base of each main hair on the leaves of this *Onosma* is surrounded by a crown of small secondary hairs, forming a second layer of spikes pointing in every direction, making it well-nigh impossible for any pest to advance further whatever its size.

Not only the leaves but also the flowers of *Phlomis bourgaei*, which grows in the garrigues of the Lycian coast of southwest Turkey, are covered in several layers of star-shaped hairs of a lovely golden colour that gives the plant a very striking appearance. These innumerable hairs end in fine points, like an inextricable tangle of spikes and barbed wire, making the plant totally inaccessible to insects. What is more, these hairs are easily detached from the leaves and are an irritant to the mucous membranes of herbivores, so that the plant is unpalatable to sheep and even goats will only browse on it as a last resort. *Phlomis bourgaei* thus gains a competitive edge in garrigues frequented by sheep and may become locally dominant in the flora of an over-grazed landscape.

Phlomis nissolii, which is one of the most beautiful silver-leaved plants in our garden, has softer but much longer tree-like hairs. These fine branching hairs are crammed together in several tiers to make a dense and particularly thick felt which gives the leaves a remarkably soft texture. Winged aphids that land on the hairy leaf surfaces of *P. nissolii* take off again rapidly as they are unable to feed: their ungainly bodies can only perch on top of the canopy of giant hairs since their rostrum, or sucking organ, is too short to reach the leaf tissue below.

• *Glandular hairs: the aromatic magic of the garrigue*

The garrigue is full of scents, especially during summer heatwaves. The gummy cistuses and butterfly lavenders of the Algarve in southern Portugal, the *Helichrysum italicum* that flows down the slopes of Cap Corse, the rue and calaminth all along the garrigue paths of Montpellier, the sages of Crete, Dalmatia and Spain (*Salvia fruticosa, S. officinalis, S. lavandulifolia*), the pink-flowered savory of the Cycladic islands (*Satureja thymbra*), the delicate marjoram of Cyprus (*Origanum majorana* var. *tenuifolium*), the myrtles, fennel and lentisks all around the Mediterranean: all these scents so characteristic of Mediterranean landscapes mingle to produce an extraordinary bouquet of aromas, turning the garrigue garden into a distillation of scent. The visual attractiveness of the garden is thus complemented by another dimension – our sensory perception of the olfactory experiences offered by garrigue plants, whose richness and complexity must be unique in the world.

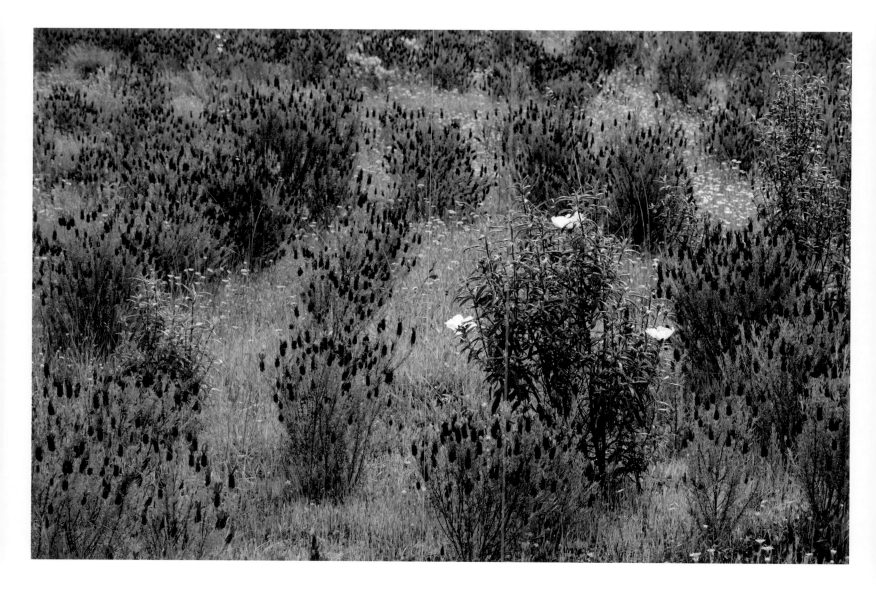

Aromatic garrigue plants release their essential oils through various organs of which the most usual are specialized hairs, the glandular trichomes. The volatile organic compounds produced by the glandular hairs help the plant to survive in a tough environment: it is the particular living conditions of the garrigue that have favoured the amazing diversity of aromatic plants that are to be found around the Mediterranean. These volatile compounds can, for example, lower the temperature around the plant, prevent the development of bacterial and cryptogamic diseases, reduce the pressure from grazing herbivores, fight off attacks by pests both directly by means of their insectifuge properties and indirectly by attracting specific insect predators, guide pollinating insects from a distance and reduce competition by inhibiting the germination of competing species through allelopathy (see p.175).

These multiple functions may be carried out by chemical variations within the same species. For example, the common thyme of the garrigue (*Thymus vulgaris*) is known to have different chemical compounds in different popu-

lations, giving rise to slight variations in scent. The chemically different types of common thyme all have their own advantages – some are more effective at repelling snails, others at repelling insects, and yet others possess allelopathic properties.

• *Sticky traps : a resin graveyard*

A species may have different types of glandular hairs according to the season, giving off scents in the garden that vary over the course of the year. *Salvia candelabrum* makes a low clump of aromatic leaves with a soft, silky texture from which in spring powerful branching inflorescences rise, every part of which – peduncles, calyces and corollas – are very sticky and highly aromatic. The undersides of the leaves are marked with bright orange beads of resin which are minuscule glandular hairs, shaped like a mushroom with a hemispheric head and attached to the leaf by a short peduncle. These glandular hairs diffuse their essential oil when a leaf is brushed against, thus breaking the cuticle that protects their secretory heads and suddenly releasing a powerful scent of sage. The volatile organic compounds metabolized by the glandular hairs have antibacterial and antifungal properties that protect the leaf. These hairs, positioned directly on the epidermis of the leaf, are surmounted by a layer of longer, silky hairs which trap a layer of aromatic air with a high concentration of the chemical compounds in close proximity to the epidermis.

The glandular hairs borne on the inflorescences are very different in appearance; they consist of a long filiform peduncle topped by a globular head which constantly exudes an extremely aromatic transparent resin that acts as a guide to pollinators. At the same time, this resin, as thick as birdlime, protects the flowers and other parts of the inflorescences from attack by insects. Any aphid misguided enough to land on the calyx of a *Salvia candelabrum* flower is trapped as if on fly-paper. Its feet are stuck fast on the sticky tops of the glandular hairs, which are

taller than the length of its rostrum, so that perched there, unable to reach the nourishing sap and equally unable to fly away, it is condemned to death. Indeed, old inflorescences of *Salvia candelabrum* gradually become graveyards full of the bodies of insects that have died of exhaustion and starvation.

The density of the glandular trichomes can vary through the seasons, so that the intensity of scents in the garden

evolves over the course of the year. The Montpellier cistus has glandular hairs that produce volatile organic compounds belonging to the terpene family, which make the plant unpalatable to sheep. The density of the glandular hairs increases in late spring and summer, pressure from herbivores becoming greater when the herbaceous vegetation surrounding the Montpellier cistuses becomes sparse with the advent of the dry period. A walk on a calm, warm evening in early summer in the coastal maquis of Corsica, with an overwhelming population of Montpellier cistuses, is an unforgettable experience: to the volatile compounds diffused by the cistuses are added the scents of *Helichrysum italicum*, germanders (*Teucrium capitatum* and *T. marum*), *Stachys glutinosa*, rosemaries, myrtles and lentisks, in an amazing bouquet of aromas. For the walker, the powerful olfactory perception of the landscape

complements the beauty of the vegetation as it takes on the colours of its summer dormancy.

• *A change of scale: when hairs reconfigure the landscape*

The chemical compounds produced by the glandular trichomes often have complementary functions. The Cretan thyme (*Thymbra capitata*, syn. *Coridothymus capitatus*) has strongly aromatic leaves, stems and flowers, the essential oil protecting the plant in over-grazed landscapes: sheep and goats won't eat it, which progressively reinforces thyme populations in the phrygana of Crete by diverting grazing pressure to other species. But the chemical compounds produced by the glandular trichomes of *Thymbra capitata* also have a powerful allelopathic action: when

Following pages: A landscape dominated by Cretan thyme (*Thymbra capitata*) thanks to the double action of the chemical compounds produced by its glandular trichomes, which both protect the plant from grazing and limit the germination of competing species.

Garrigue plants can be seen from different and complementary viewpoints.

1. The biologist's viewpoint: observation of a phlomis leaf (*Phlomis bourgaei*) under a stereoscopic microscope allows one to see the structure of the thick mattress of trichomes that cover the epidermis. The trichomes limit water loss while at the same time protecting the plant from herbivores.

2. The gardener's viewpoint: phlomises are of interest in the garrigue garden both for their massive flowering and for the winter structure of their velvety foliage. This scene includes *Phlomis* 'Le Chat' (which is a hybrid of *Phlomis purpurea* and *Phlomis crinita*), *Lomelosia minoana*, *Euphorbia ceratocarpa* and *Teucrium fruticans* 'Agadir'.

3. The ecologist's and landscape designer's viewpoint, which studies the dynamics of the evolution of the ecosystem: self-seeding of *Phlomis fruticosa* in the ruined walls of former fields, among other pioneer plants like tree euphorbias (*Euphorbia dendroides*) and Phoenician junipers (*Juniperus phoenicea*).

they are diffused in the soil from fallen leaves or after the foliage has been washed by rain they inhibit the germination of competing species. Thus the production of chemical compounds benefits the plant twice over, by an adaptation to grazing and by fighting off other species potentially adapted to the same environment. In this way one may sometimes see in Crete scenes of great beauty where this thyme, remarkable both for its long summer flowering period and for its perfectly regular cushion shape, has become the dominant plant – the structure and composition of the landscape as far as the eye can see is directly linked to the almost invisible hairs on the thyme. In the setting of a garrigue garden, the use of aromatic plants such as the Cretan thyme or countless other species with glandular trichomes that produce aromatic compounds, including rosemaries, oreganos, sages, micromerias, savories, lavenders, santolinas and helichrysums, enables one to envisage a different way of managing the garden where weeding is reduced thanks to the amazing allelopathic potential of essential oils (see page 175).

The diversity of textures in the foliage of garrigue plants can be looked at from several complementary viewpoints: that of the biologist, for whom extraordinary miniature landscapes open up on looking through a microscope or steroscopic microscope, where the appearance of the epidermis, the arrangement of droplets of essential oil and the shape of the trichomes express the fascinating strategies that enable plants to withstand tough living conditions; that of the gardener, for whom the beauty of the foliage is set off by a play of contrasts between shapes, colours and textures, which, together with the way plants catch the light and the diffusion of their scents, give an infinite range of possibilities and offer an inexhaustible source of inspiration for planning new scenes in the garden; and that of the landscape designer, whose viewpoint may meet that of the ecologist in analysing the dynamics of the vegetation, whether in the garrigue or in a garden inspired by the garrigue, where the diversity of foliage expresses specific adaptations to the environment that confer competitive advantages on some species, both on the scale of the wider landscape and on the smaller scale of the garden. The garrigue garden allows these various viewpoints to converge, inviting the gardener to expand his or her view and not only consider how a diversity of foliage contributes to the beauty of the garden throughout the year but also envisage a new way of maintenance, one that exploits the strategies inherent in plants to manage pests and reduce weeding. In a garrigue garden, the immediate visual beauty that is provided by a diversity of foliage thus reflects an underlying 'ecological beauty' that enables us from the very first planning stage to integrate a future management plan for the garden based on reduced maintenance.

THE RHYTHM OF THE LANDSCAPE

Page 122: Landscape sculpted by goats near the Omalos plateau in Crete: *Quercus coccifera*, *Acer sempervirens* and *Zelkova abelicea*.

Top: A flow of silver balls on the black sand of Etna with *Astracantha sicula* in the foreground.

Bottom: The sphere is the shape that has the smallest surface area for a given volume. *Rhamnus lycioides* subsp. *oleoides* maintains a high level of photosynthesis thanks to its hemispheric habit: its volume of foliage remains significant although the plant exposes the smallest possible surface to wind, salt spray and repeated attack by herbivores.

Walking on the slopes of Mount Etna is a thrilling experience. The path leading up to the summit of the volcano crosses an austere landscape of black or grey scoria. When you reach the crater zone, at more than 3000m (9843ft), the heat from the fumeroles and the smell of sulphur are almost suffocating. No vegetation manages to grow near the summit, but as you make your way back down there are some lovely surprises: if you leave the main path and cut across the steepest slopes, your feet slithering in the volcanic sand as if it were powdery snow, you may discover the first colonies of pioneer plants clinging to the side of the volcano. Sicilian astragalus (*Astracantha sicula*), Sicilian soapwort (*Saponaria sicula*), Etna bedstraw (*Galium aetnicum*) and Sicilian tansy (*Tanacetum siculum*) form flows of grey or silver balls hugging the black sand. In late spring, when temperature and humidity conditions are the most favourable, these cushions are suddenly covered in colour, their pink, yellow, white or red flowers using bright hues the better to attract the few pollinators that exist in this almost entirely mineral land-scape. But it is in late summer that the Etna landscape becomes most beautiful, when the flowering season is over. The simple contrast between the rounded silhouettes of silver cushions and the expanses of black lava lying in folds makes an utterly striking picture, with in the background the line of the coast stretching in the distance to the town of Syracuse.

The pioneer plants that grow high up on the slopes of Etna have evolved to survive in an extreme environment. They need to tolerate cold and snow in winter as well as brutally high temperatures in summer, when the dark volcanic substrate becomes burning hot under the rays of the sun and abrasive grains of sand are lifted by the gusty wind that swirls around the peaks of the volcano. In order to withstand these conditions, the various species have adopted a common strategy, which is what gives this land-scape its unique character: they all grow as cushions or balls, their vegetation seeking to come as near as possible to a perfect sphere. A spherical shape is an adaptation in plants to a hostile environment. For a long time ancient mathematicians were intrigued by the sphere but were unable to measure its volume or surface area. It was indeed in Syracuse that the Greek scientist Archimedes was the first person to penetrate the mystery of the sphere. In discovering how to calculate its volume and surface area, he showed that it possesses a property that distinguishes it from all other geometric figures: the sphere is the shape that has the smallest surface area for a given volume. It is this relationship between volume and surface area that explains the rounded form of so many garrigue sub-shrubs living in harsh conditions: in order to withstand the stresses of their environment better, they huddle in on themselves and take on a ball- or cushion-shaped habit. By adopting a spherical or hemispherical shape they optimize growth and maintain a high rate of photosynthesis thanks to their large foliage volume, while at the same time exposing the smallest surface possible to heat, cold, high solar radiation, wind, the abrasive action of sand, scorching by salt spray and repeated assault by herbivores.

The ball shape is such an effective resistance strategy that all around the Mediterranean we see garrigue landscapes with a beauty that comes, as on the slopes of Etna, from the rhythm created by a succession of plants with compact and rounded shapes. All along the coast of Cap Corse, the undulating vegetation of the coastal maquis is composed of cushions of rosemary, cistuses (*Cistus salviifolius*), germanders (*Teucrium capitatum*) and *Stachys glutinosa*. On the footslopes of the M'Goun mountain in the Atlas, in Morocco, the landscape is given rhythm by the mighty rounded masses of succulent euphorbias (*Euphorbia resinifera*) along with balls of lavender (*Lavandula dentata*),

polygalas (*Polygala balansae*) and shrubby globularias (*Globularia alypum*). In Mallorca, on the Levant peninsula, silvery cushions of santolina (*Santolina magonica*) and grey balls of *Helichrysum italicum* colonize the stony expanses swept by spindrift. In the dunes around the bay of Balos in Crete, massive balls of heather (*Erica manipuliflora*) and thyme (*Thymbra capitata* syn. *Coridothymus capitatus*) create a landscape that looks entirely mineral as the plants have become so hard and tight they resemble rocks embedded in the sand. Near Cape Saint Vincent in southern Portugal, bright pink thyme (*Thymus camphoratus*), brilliant yellow broom (*Stauracanthus spectabilis*),

In the Balos dunes in Crete, heather (*Erica manipuliflora*) that turns reddish-brown after flowering and silver thyme (*Thymbra capitata*) create a landscape that looks entirely mineral as the plants become so hard and tightly packed that they resemble rocks half-buried in the sand.

A landscape of balls and cushions at Cape Saint Vincent in southern Portugal, where the white flowers of *Cistus ladanifer* var. *sulcatus* mingle with the pink flowers of *Armeria pungens* and the yellows of the broom *Stauracanthus genistoides* and the daisy-flowered *Pallenis maritima*.

cistuses with large white flowers (*Cistus ladanifer* var. *sulcatus*), pink thrift (*Armeria pungens*) and mauve lavenders (*Lavandula pedunculata*) compose an extraordinary landscape in which the coloured balls form waves of vegetation that continue to the top of the cliffs overlooking the sea.

• Landscapes of light

The strategies plants use to adapt to environmental conditions lead to a repetition of shapes, with species belonging to completely different families sometimes having such a similar appearance that it can be hard to tell them apart when they are not in flower. This repetition of shapes makes for a remarkable landscape architecture where the rhythm of the light reinforces the sense of depth. Landscapes consisting of cushions and balls are sculpted in a spectacular manner by the way they are lit on their sunny and shaded sides. A range of simple shapes appears whose beauty comes across equally well at different scales, from the miniature in a few square metres to the large in landscapes that sometimes stretch as far as the eye can see.

The beauty of these landscapes of light may serve as an inspiration for gardeners, helping to make the garrigue garden attractive throughout the year. In the mediterranean climate a garden's appearance changes profoundly from season to season and may often seem less interesting in summer when the profuse spring flowering is over, yet the garrigue garden can be strikingly attractive during the period of summer dormancy. In spring the exuberance of colour makes one almost forget the underlying structure of the garden, but at the height of summer, when fewer plants are in flower, the structure of the garden regains its importance. Thanks to the numerous garrigue plants with a natural habit that is a cushion or ball shape, the rhythm of the light throws silhouettes into relief, accentuates contrasts and chisels textures. Leaving behind the traditional palette of summer flower colours, the garden acquires a different kind of visual power, rather like that of black and white photographs; at the height of summer, modelled in countless nuances of grey, silver and dark foliage, the garrigue garden becomes an exceptional landscape of light.

1. The beauty of landscapes consisting of cushions and balls can be seen on different scales, from a few square metres to expanses stretching as far as the eye can see. *Santolina magonica* on the Aromatic Path in the garden of the Fort Saint-Jean looking out over Marseilles.

2. A landscape of light: *Thymbra* and *Ballota acetabulosa* in Crete.

3. *Centaurea spinosa* forms a miniature garden by the side of the road on the Cycladic island of Sifnos.

4. In spring the riot of colour almost makes us forget the underlying structure of the garden. Nevertheless this structure is fundamental to the visual attraction of the garden when plants are not in flower.

Following pages: Many landscapes in the Cyclades, exposed to wind, salt, drought and grazing, are dominated by ball- and cushion-shaped plants: *Centaurea spinosa* by the sea on the island of Kea.

On the island of Antiparos in the Cyclades, the remarkable gardens created by Thomas Doxiadis are conceived as magnificent landscapes of light. In an environment where the arid hillsides are entirely covered all the way down to the sea with plants growing in tight balls – *Centaurea spinosa*, *Convolvulus oleifolius*, *Sarcopoterium spinosum*, *Teucrium brevifolium* and *Thymbra capitata* – his gardens opening directly on to nature consist of plants that not only have the same compact and rounded shapes but also the same colour palette of grey and green tones, ensuring a perfect continuity in the rhythm of light which is here the link between the garden and its surrounding landscape.

In summer, the garrigue garden best reveals the beauty of its architecture in the early morning and evening, when the low light accentuates shadows and brings the rhythm of the plants to life. In the Mediterranean, the garden is less of a draw on hot afternoons when the light is overhead, blurring shapes and obliterating the visual appeal of the structure. In these hottest hours of the day, interest lies more in the scents of the plants than in their visual appearance, the essential oils of thymes, oreganos, sages, helichrysums, rues, santolinas, lavenders, artemisias and cistuses mingling in the powerful aromatic bouquet that is so characteristic of garrigue gardens. In winter, by contrast, when the foliage is less aromatic and the sun's rays are oblique throughout the day, the light sculpting the landscape transforms the garrigue garden into a vast living kaleidoscope, with the changing angles in the play of light and shade revealing new perspectives all day long.

At the Fort Saint-Jean in Marseilles, the concept of the garden is based on the combined effect of the summer scents of the plants and the play of light and shade created by the constant structure of evergreen foliage. The Aromatic Path that overlooks the outlet channel of the Old Port, for example, presents us with a succession of cushion-shaped plants, the beauty of which can be appreciated as much at a distance as close to: seen from afar, for instance from the Belvedere des Figuiers which dominates the old sentry walk, the rhythm of the plants' supple shapes is reinforced by the contrast with the powerful

architecture of the Fort's walls, while from close quarters, when one can gently stroke leaves to discover their scents, the regular shapes of nepetas, santolinas and helichrysums form waves of silver against the backdrop of the sea.

Inspired by the garrigue, gardens of cushion- and ball-shaped plants are dynamic landscapes where the play of light and shade changes over the seasons and can also be modified from year to year. In one section of our garden we have created a scene dominated by low-growing compact and rounded plants which recall the landscapes of Corsica, Mallorca and Crete. The main framework is provided by a wide botanical palette, with collections of phlomises, artemisias, cistuses, ballotas, euphorbias, santolinas, germanders and sages, the different species within each genus being chosen for the similarity of their shapes.

Other species, planted in a more one-off way, including *Ptilostemon chamaepeuce*, *Ebenus cretica*, *Cneorum tricoccon*, *Lomelosia hymettia*, *Bupleurum spinosum* and *Dorystaechas hastata*, form a secondary and more discreet framework which enriches the scene by the textures or striking colours of the foliage. All these plants have shapes that are sufficiently alike for their positioning to be incidental: in this area of the garden we chose to plant them at random, without following a planting plan.

Most cushion- or ball-shaped plants from garrigue landscapes are pioneer plants that self-seed easily in the garden. We make the most of this ability to self-seed, allowing the position of the cushions of sages, ballotas, phlomises and cistuses to alter progressively, while preserving the same frame of light and shade, at once evolving and constant.

• A landscape in two layers

The village of Biniaraix in Mallorca is situated in the heart of the cultural landscape of the Serra de Tramuntana. Listed by UNESCO as a World Heritage Site, this unique landscape comprises a network of paved paths connecting the countless terraces, some of them tiny, supported by carefully maintained dry-stone walls, as well as a complex irrigation system dating from the time of Moorish rule (902–1230CE). The combination of underground channels, holding basins and distribution ditches has made it possible over the centuries to manage a mosaic of cultivated plots won from the rocky slopes. Near the village, vegetable gardens alternate with orchards of citrus and other fruit trees. Higher up, as the land becomes poorer, the citrus trees are progressively replaced by terraces of figs,

Top: In summer, it is in the mornings and evenings that the garrigue garden reveals the full beauty of its architecture, when the low light emphasizes shadows and brings to life the rhythm of the vegetation.

Bottom: The Aromatic Path in the Fort Saint-Jean in Marseilles: a succession of scented plants positioned so that visitors can touch them.

Top: Scents and architecture in the garden: helichrysums, rosemaries, sages and lavenders at the Fort Saint-Jean in Marseilles. The concept for the garden is drawn from the way the summer scents of garrigue plants complement the play of light and shadow created by the architecture of evergreen foliage.

Bottom left: Most ball- or cushion-shaped plants that come from garrigues are pioneer plants which self-seed easily on stony ground. *Santolina villosa* colonizes a gravelled former parking lot near Morella in north-east Spain.

Bottom right: In this area of our garden we have achieved a random planting: the positions of the plants change from year to year according to the way phlomises, cistuses, ballotas, sages and euphorbias self-seed.

almonds, olives and carobs, the twisted trunks anchored in the walls recalling the long history of human labour, intimately linked with the history of the landscape.

The terraces furthest from the village, which are accessible only on foot or by riding a mule along narrow stone-paved paths, have today mostly been abandoned. The garrigue lies in wait for abandoned plots of land and spontaneous

plants seed themselves at the foot of the trees. Euphorbias (*Euphorbia characias*), asphodels (*Asphodelus microcarpus*) and *Cneorum tricoccon* were the first plants to establish themselves, followed by myrtles, lentisks, phillyrea and buckthorn. The higher one climbs the Biniaraix valley on the main path that snakes between the terraces up to the summit of Puig de l'Ofre, the more wild species can be seen to have invaded cultivated land, until little by little

garrigue plants make garden-like scenes between the trees. The supple clumps of *Ampelodesmos* form borders along the paths and *Clematis cirrhosa* tumbles from the walls, while beneath the olive trees rosemary, helichrysums, germanders and globularias intermingle with cistuses and heathers. In the spring, a multitude of wild flowers – orchids, narcissi, gladioli, ornithogalums and grape hyacinths – emerge from the cracks between the stones. When summer comes, fennel, calaminth, rue and pitch trefoil release their powerful scents as soon as a passerby brushes against their leaves. Here the garrigue landscape is gradually taking over agricultural land and the abandoned old plots are slowly being transformed into gardens where the rhythm of the landscape is expressed in two dimensions, horizontal and vertical. The compact bushes of myrtle, lentisk, heather, helichrysum and cistus create the horizontal rhythm, spreading from terrace to terrace, while the vertical rhythm derives from the contrast between the low layer of garrigue plants and the taller layer of the trees that give structure to the landscape all along the valley.

This two-layered structure characterizes the evolution today of many Mediterranean landscapes that derive from an ancient tradition of agroforestry. On the island of Sifnos in the Cyclades, the mule path that links the villages of Apollonia and Kastro crosses a cultivated valley, punctuated by tiny monasteries, where vegetable plots marked by cypress trees alternate with strips of cereal crops grown beneath the olives. On Sifnos agricultural activities have

to a large extent been replaced by tourism, and garrigue plants exploit this by propagating freely in the cultivated fields. Tree euphorbias (*Euphorbia dendroides*), pink savory (*Satureja thymbra*), silver ballotas and white-flowered oregano (*Origanum onites*) are in the vanguard as the phrygana advances, seeding themselves everywhere beneath the olives and along the stone walls. In the early summer light, when the cereal crops are golden and the tree euphorbias take on flamboyant hues of red, orange and mauve, the contrast of colours and the double-layered rhythm of rounded euphorbias, old olive trees and dark rows of cypresses remind us of how tenuous the border between a human landscape and a natural garden can be in these landscapes that have been worked by man since antiquity.

In other agroforestry landscapes too, subject to a recent transformation, the lower layer is home to a profusion of pioneer garrigue shrubs and sub-shrubs. On the Noto plateau in Sicily, the stony land planted with carobs is being invaded by sages (*Salvia fruticosa*) and *Prasium majus*; in the valleys of the Anti-Atlas, the terraced plots of barley beneath almond trees are being colonized by dark junipers (*Juniperus thurifera*) and white broom (*Retama monosperma*); in the Alentejo in southern Portugal, the montados, those vast traditional landscapes where cork oaks are scattered throughout the pastures on the hills, are being inexorably invaded by gum cistuses (*Cistus ladanifer*), crispus rock roses (*C. crispus*), sage-leaved cistuses (*C. salviifolius*) and Spanish lavender (*Lavandula stoechas*).

For Mediterranean gardeners, the inspiration drawn from the meeting between two landscape models linked to human history – garrigue landscapes and agroforestry landscapes – is unbeatable. The two-layered structure makes it possible to adapt a garrigue garden to many different situations thanks to its aesthetic and functional qualities: it provides at the same time the shade so sought after in the summer garden and a groundcover that reduces maintenance throughout the year. The gardens created in Greece by the landscape designers Jennifer Gay and Piers

Bottom right: Abandoned agroforestry landscape near Tiznit in southern Morocco: two layers of vegetation are created as King Juba's euphorbias (*Euphorbia regis-jubae*) colonize the ground beneath argan trees.

Opposite: Old trees form the structural framework of the gardens created by Jennifer Gay and Piers Goldson which fill all the empty spaces in the village of Rou in Corfu, restored by the architect Dominic Skinner.

Goldson illustrate the possibilities of the two-layered rhythm in very varied environments. At Rou, an old mountain village in Corfu transformed into a tourist complex, the gardens give unity to the restored village, planted along the lanes and narrow passages between the houses, in tiny squares and internal courtyards. The management of this landscape began by using judicious pruning to show off to advantage the tree heritage of the abandoned village, which consisted of a mixture of old trees initially planted to provide food, shade or ornament (almonds, figs, pomegranates, olives, mulberries, arbutuses, plane trees, Judas trees and cypresses) and the spontaneous vegetation that had taken over the spaces left empty by humans (holm oaks, phillyreas and turpentine trees). The second stage consisted of restoring all the walls, steps, terraces and paved paths, planting them with a low layer of cushion-shaped, ball-shaped and carpeting plants, using evergreen species to fill the spaces between the hard landscaping in order to reduce the work of weeding in the multitude of little patches interwoven among the buildings. To echo the intermingling of spontaneous and planted trees, the plant palette for the lower layer mixes cultivated and wild species, with aromatic and ornamental plants such as artemisias, rosemaries, savories, oreganos, lavenders and sages growing in association with wild plants that self-seed spontaneously in different parts of the gardens, such as cistuses, lentisks and buckthorns. The

smallest free spaces are planted in a way that mixes the mineral and the vegetable: old walls are transformed into planters for thymes, irises and sedums, stone benches built in the shade of old trees are surrounded by acanthuses, steps are transformed into cascades of valerian, Cupid's dart and asphodels. These multi-faceted double-layered gardens at Rou invite visitors to take a walk of botanic discovery from the village to the surrounding garrigue.

In the Argolid Peninsula of the Peloponnese, Jennifer Gay and Piers Goldson have designed a garden in a very different environment, on a site right by the sea. Here the principal constraints are the wind and salt spray as the meltemi, the wind that blows in summer, regularly drenches the Aegean coast with salt. A grid of pathways was created beneath the existing tree layer of pines and eucalypts, linking the garden to the coastal path. The plants between these pathways were chosen for their ability to resist the harsh conditions of the coast, many cushion- or ball-shaped plants being naturally adapted to such a life. Phlomises, ballotas, cistuses, rosemaries, cinerarias, helichrysums and silver-leaved convolvulus cover the ground and form the lower layer of this seaside garden, the plants in the first line protecting those behind them, as in the natural structure of coastal flora where successive sheets of vegetation protect each other from the pressures of wind and salt spray.

One of the most beautiful gardens created by Gay and Goldson is that of Vilka and George Agouridis near Athens. This terraced garden looks out over the agricultural plain of Attica, where olive groves are surrounded by abandoned plots colonized by a young phrygana in which the silver, russet and golden colours of centaureas

Following pages: The transformation of a former agroforestry landscape partially overrun by tree euphorbias, in the valley between Apollonia and Kastro on the Cycladic island of Sifnos.

A double-layered garden in the Peloponnese: the lower layer consists of undulating waves of *Limoniastrum monopetalum*, *Phlomis chrysophylla*, *Origanum syriacum*, *Helichrysum italicum*, *Cistus albidus*, *Rosmarinus officinalis* 'Boule', *Senecio cineraria* and *Teucrium fruticans*.

Opposite: A double-layered garden inspired by old agroforestry landscapes around the Mediterranean: the structural trees are complemented on the ground by a low-growing layer of grey, green and silver masses of artemisia, ballota, sage, phlomis and helichrysum. The garden of Vilka and George Agouridis, designed by Jennifer Gay and Piers Goldson.

(*Centaurea spinosa*), spiny burnet (*Sarcopoterium spinosum*) and Jerusalem sage (*Phlomis fruticosa*) are blended. The upper terrace, which one comes upon first when one enters the garden, presents a remarkable fusion of the key elements of the surrounding landscape: the strong structure of the old olive trees which existed on the property before the house was built, and a lower layer of new garrigue planting, arranged with finesse and evoking the colours of the adjacent phrygana. On either side of the path that leads the eye down the terrace, the rhythm of the landscape is expressed in a succession of balls with silver, grey or golden leaves set against the dark shapes of the olive trunks. The success of this area of the garden is due in part to the richness of the plant palette used and to the diversity of leaf textures in the lower layer. Artemisias (*Artemisia arborescens* and *A. thuscula*), phlomises (*Phlomis chrysophylla*, *P. italica*, *P. bourgaei*, *P. purpurea*, *P. lycia*), cistuses (*Cistus creticus* and *C.* × *pauranthus*) and ballotas (*Ballota pseudodictamnus*, *B. acetabulosa*, *B. hirsuta*) are complemented by phillyreas, arbutuses, germanders, *Vitex agnus-castus*, coronillas, myrtles, atriplex (*Atriplex halimus* and *A. canescens*) and bupleurums, the taller shrubs forming a mixed hedge along the upper wall of the garden. The lower terraces, planted with a collection of fruit trees, go up to the house; the microclimate in the sheltered lee of the south side of the house makes it possible to cultivate here a small garden oasis of citrus trees surrounded by cypresses, almond trees, pomegranates

and olives. A gravel path, partially invaded by carpeting groundcover plants, *Artemisia pedemontana*, *Achillea crithmifolia*, *Frankenia laevis*, *Psephellus bellus* (syn. *Centaurea bella*) and *Achillea umbellata*, threads its way between the trees. It creates the transition to an area of low cushion-shaped plants, *Artemisia alba*, *Lavandula* × *ginginsii*, *Helichrysum orientale*, *Lomelosia cretica* and *Salvia lavandulifolia*, surrounded by an edging of prostrate rosemary, buzzing with bees in spring, that flows over the walls above the other fruit trees.

The two-layered structure is used by many landscape designers around the Mediterranean. In the hills above Saint Tropez, the garden of the Hotel Muse, the work of the landscape designer Sophie Ambroise, consists of a series of curving paths beneath a cover of almonds and olive trees. The sides of the paths are planted with groundcover in shades of grey-green, blue-grey and silver; the carpets of thyme, glaucous grasses, grey achilleas and artemisias with finely cut leaves have naturalized to look like a silvery steppe growing beneath the trees. In Mallorca, the garden of Camilla Chandon, hidden away like a secret garden at the foot of a cliff at the end of the Serra de Tramuntana, has a spectacular two-layered structure where carobs, cypresses and almond trees emerge from a low-growing garrigue vegetation, in which the rounded masses of rosemaries, viburnums, myrtles and lentisks are carefully clipped by such a meticulous gardener that their uncluttered style evokes a zen-inspired garden.

On a different scale, the park of the Stavros Niarchos Cultural Centre in Athens, designed by the landscape architects Deborah Nevins and Elli Pangalou, is laid out like a two-layered garrigue garden on an area of more than 10 hectares (25 acres). Rows of oaks, plane trees, carobs, figs and olive trees are aligned along the broad pedestrian paths that demarcate the plantings. Conceived as an evocation of Mediterranean vegetation, the park has an extensive palette of garrigue plants: euphorbias (*Euphorbia ceratocarpa*, *E. dendroides*, *E. myrsinites* and *E. spinosa*)

are blended with numerous varieties of rosemary, sage, artemisia, cistus, lavender, St John's wort, helichrysum and phlomis, in a multitude of combinations which form flows of grey, green and silver foliage beneath the trees. This garrigue garden then gives way to an immense steppe of supple golden grasses on the spectacular green roof on top of the new Athens Opera House.

• *Spatial rhythm and the dynamics of evolution*

The landscape designer Sandrine Lefèvre has created a striking landscape arrangement in the new extension to the cemetery of Saint-Marc-de-Jaumegarde, situated at the foot of the imposing Mont Sainte-Victoire, near Aix-en-Provence. In the main section of the new cemetery, the spaces between the graves are planted with a mixture of allelopathic groundcover plants (see p.175) which reduce the amount of maintenance required, part of an initiative to lessen the use of chemical herbicides in the public spaces of the village. A garden for strolling and meditation provides a transition between the cemetery and the surrounding pine forests. Between the young trees – oaks, maples and olives – a low-growing layer of helichrysums (*Helichrysum stoechas* and *H. orientale*), othonopsis (*Hertia cheirifolia*), goniolimons (*Goniolimon speciosum*) and euphorbias (*Euphorbia rigida*, *E. characias* subsp. *wulfenii* and *E. × martini*) has been planted on a bed of gravel. The prolific self-seeding of the euphorbias into the gravel from the very first year after they were planted augurs well for the evolution of this newly created garden.

Top: Green paths beneath almond and olive trees. Garden designed by Sophie Ambroise for the Hotel Muse at Saint Tropez.

Bottom: Rosemary clipped into topiaries beneath almonds and cypresses. Garden designed by Camilla Chandon in Mallorca.

The dynamics of evolution are an integral part of garrigue gardens and many Mediterranean landscape designers now envisage right from their first plans the progressive transformation of the space. Instead of defining a precise picture corresponding to the final culmination of the garden, they seek simply to put in place the elements of an evolution that will pass through several stages, all of them ornamental but looking very different depending on the age of the garden. The future development of it cannot be exactly determined, since the plant dynamics in each garden vary according to the local weather and soil conditions.

Claudia and Udo Schwarzer, landscape designers and ecologists who work in southern Portugal, have been studying the rhythm with which garrigue plants evolve for a long time. Their own garden is situated at the foot of the Serra Monchique, known to botanists for being the home of populations of rare native plants such as Azores myrica (*Myrica faya*), Monchique euphorbia (*Euphorbia paniculata* subsp. *monchiquensis*) and *Rhododendron ponticum*, which benefit from this mountain's mild and relatively damp climate due to its proximity to the ocean. (British readers will know that the climate conditions of the UK suit *Rhododendron ponticum* very well, which is why, having originally been imported as an ornamental garden plant, it has now become naturalized and indeed invasive.) The history of their garden began with the pur-

chase of a 5 hectare (12 acre) plot – an old eucalyptus plantation such as one frequently sees in Portugal. After first of all uprooting the eucalypts, causing a major disturbance of the soil when the stumps were removed, the Schwarzers left the spontaneous flora to evolve by itself, having decided to treat their garden as a laboratory in which to observe the natural dynamics of the vegetation.

During the first years, the disturbed soil was colonized by a rich array of the annual, biennial and perennial plants that one finds in abandoned fields in the Algarve, with spectacular and colourful spring flowering: the vibrant blue of viper's bugloss (*Echium vulgare*), the bright pink of campions (*Silene colorata*), the white of chamomiles (*Chamaemelum mixtum*) and the brilliant yellow of

Claudia and Udo Schwarzer study the dynamics of the evolution in the vegetation of their garden in Portugal: here young arbutuses give structure to the space among cistuses and lavenders.

spotted rock roses (*Tuberaria guttata*). Later on, a young garrigue dominated by cistuses and lavenders (*Lavandula stoechas* and *L. viridis*) took over, with the layer of evergreen sub-shrubs gradually replacing the herbaceous plants. Then, after about ten years, the garden began to acquire structure, with layers of plants of different heights as woody species such as buckthorn, phillyrea, lentisk and arbutus, mixed with young oaks, appeared amid the cistuses and lavenders. The oaks subsequently emerged from the garrigue, with the garden reaching a first stage of maturity, characterized by great botanical richness, after about 15 years.

Today, narrow paths allow one to walk through the exuberant garrigue in this garden of multiple layers consisting of shrubs, oak trees and clearings – for Claudia and Udo Schwarzer maintain open areas at several points in the garden in order to preserve the diversity of wild herbaceous plants. As often happens in southern Portugal, the different species of cistus hybridize freely, resulting in one of the most remarkable natural collections of cistuses that we have ever seen in a garden. Among the species are *Cistus inflatus*, with slightly hairy leaves and white flowers; *C. ladanifer*, with sticky leaves and large blotched flowers; *C. monspeliensis*, with dark leaves and small white flowers; *C. crispus*, with wavy leaves and bright mauve-pink flowers; *C. albidus*, with felty grey-green leaves and pale pink

flowers; and *C. salviifolius*, with rounded leaves and off-white petals surrounding a large central bunch of yellow stamens. Since cistus hybrids often inherit the qualities of both their parents, in them the garrigue offers gardeners particularly robust and floriferous plants. Claudia and Udo Schwarzer have been able to observe the spontaneous appearance of some remarkable cistuses, among them *C. × pulverulentus*, an especially floriferous cross between *C. crispus* and *C. albidus*, and the vigorous *C. × verguinii*, a cross between *C. salviifolius* and *C. ladanifer*, reputed to be one of the cistuses that live longest in gardens.

Near the Île-Rousse in Corsica, the Park of Saleccia, created by Bruno and Irène Demoustier, is organized entirely around the rhythm created by the evergreen foliage of lentisks, myrtles, phillyreas, oleasters, buckthorns, bupleurums, rosemaries, heathers, cistuses, euphorbias and helichrysums, which provide the main structure of the vegetation. Balagne is a region of Corsica famed for its magnificent light. The Park of Saleccia, the architecture of which is without doubt one of the most accomplished in the Mediterranean, is designed as an ode to light in every minute detail, through the contrasts, shapes and textures of maquis species and the juxtaposition of plants left to grow freely with plants clipped to various heights, creating a spectacular effect in this huge 7-hectare (17-acre) garden. As with the garden of Claudia and Udo Schwarzer, the origins of the Park of Saleccia lie in a major initial disturbance, in this case a violent fire which in 1974 ravaged an old olive grove in a valley facing the sea. Following the fire, the restoration of the Park of Saleccia became the story of the tireless labour of a passionate gardener, who has worked for several decades to create a remarkable garden simply by guiding the natural reconquest of the site by maquis vegetation. After grafting the olives which were growing again from their rootstocks after the fire, Bruno Demoustier patiently shaped some of the spontaneous plants that seeded themselves under the trees by regular clipping, leaving other plants to grow freely according to their natural habit. The dark silhou-

Designed by Bruno and Irène Demoustier, the Park of Saleccia is the story of a passionate gardener who has worked for several decades to create a remarkable garden, guiding the natural re-establishment of the maquis vegetation after a fire in an old olive grove.

ettes of sclerophyllous shrubs, the olives once more resplendent over the whole hill, their foliage silvery when tossed by the wind, the silver expanses of helichrysums and euphorbias and the occasional counterpoint of cypresses and oaks now form the garden's main structure. Over the years, Bruno and Irène Demoustier have enriched the palette of maquis plants with a large collection of plants adapted to the conditions of Balagne deriving from other countries around the Mediterranean as well as from other regions with a mediterranean climate, making a demonstration garden that today is one of the most visited gardens in Corsica.

The Park is subdivided into many different gardens: the evergreen shrubs create a network of interwoven screens which allow one to discover little by little all the gardens that open up successively as one walks through the olives. Each piece of this complex mosaic has its own atmosphere deriving from the choice and arrangement of species, the play of light on their various textures and foliages, and the inventive rhythm, never the same, between trimmed and free-growing plants. The Park of Saleccia has now reached its full maturity, yet always changes from year to year since gardens inspired by the garrigue are in constant evolution. In the nursery adjoining the Park, Bruno Demoustier propagates new species every year, always seeking unexpected plant associations that will further refine the subtle plays of light and shade which unite the different areas of the Park.

• Anticipating the rhythm of evolution in a garden

In the wild, the trajectories of the evolution of garrigue landscapes depend on the external forces that influence their plant dynamics. In the same way, the garrigue species included in a new garden planting may develop in different directions. When planning a garrigue garden it is possible to anticipate its future transformation by creating distinct zones where phases will follow different time scales. Depending on the choice of plants, the density at which they are planted, the quality of the drainage and

Top: The Park of Saleccia was conceived as an ode to light, on the spectacular scale of a 7-hectare (17-acre) garden.

Bottom: Evergreen shrubs make a network of interlinked dividers which allow discovery little by little of the successive gardens that open up between the trees.

Good news for impatient gardeners: the main layer of a garrigue garden, consisting of a variety of sub-shrubs from around the Mediterranean, grows fast. The garden of the Fort Saint-Jean in Marseilles four years after planting.

the type of mulch, inorganic or organic, we can steer the type of evolution and the speed with which a particular zone of the garden will reach the degree of maturity that corresponds to the result we want, while at the same time requiring little maintenance.

Novice gardeners are often impatient. They hope to achieve an outside space that will look like a mature garden as rapidly as possible. The good news for gardeners in a hurry is that the principal layer of a garrigue garden, comprising various sub-shrubs from all around the Mediterranean Basin, grows very fast. A young garden consisting of cistuses, artemisias, rosemaries, phlomises, santolinas and sages can look like a mature garden after only a short time, about two to five years after planting.

Heliophile sub-shrubs from the garrigue in fact seek to colonize as fast as possible any spaces left open by any disturbance of the environment – and to plants, the work we carry out when we are making a garden seems just like the disturbances they might encounter in the wild. Thus the gardener's timescale and that of the garrigue may find themselves in harmony, a garrigue garden taking shape very rapidly in its initial growth phase. Recently created garrigue gardens like those of the Fort Saint-Jean in Marseilles or the Stavros Niarchos Park in Athens demonstrate this: only a few years after they were planted, these much-visited urban spaces are perceived by the public as mature gardens although in fact they are still very young and have been designed to evolve even more strongly in the future.

The speed with which a garden can be achieved, the direction it takes and the maintenance needed to keep it in line with the gardener's expectations in the long term can all be defined at the planning stage. In our garden we have varied the dynamics of evolution in different ways by creating distinct zones. Some zones are free; they provide us with a lot of information and for the last 20 years or so have allowed us to observe the behaviour of a garrigue flora left to itself in our local climate and soil conditions. Other zones are partially controlled, with selective annual weeding of excess seedlings to hold back the ebullience of pioneer shrubs such as coronillas, bupleurums and viburnums. There are also zones that are conceived of as landscapes in equilibrium; here the use of an inorganic mulch and the dominance of allelopathic plants (see p.175) in the form of carpets, cushions or balls make it possible to maintain a stable vegetation structure over the long term.

Our garden began with the kind of disturbance that is classic in the history of many Mediterranean gardens located in agricultural areas. At the beginning it was a not particularly attractive piece of wasteland left after an old vineyard had been uprooted, flat and with calcareous clay soil compacted by building work and the passage of heavy machinery. We created the different zones in our garden in stages. It took several years for the whole site to be developed, since the labour involved in the preparatory work – constructing low walls, laying out paths, raising planting areas and sometimes adding sand and gravel to improve drainage – obliged us to develop it section by section. The maintenance of the different zones was, over time, an apprenticeship in the behaviour of the collection of garrigue plants that we had put together; indeed, from the beginning the garden was an experiment to enable us to understand better how the dynamics of the plants could be harnessed to reduce maintenance in gardens.

The various zones in our garden have become notably differentiated with the passage of time. In the first years after planting most of the beds looked relatively similar.

To improve the rate of establishment and reduce the need for water after the first year, we chose to plant only small specimens, regardless of their future size. Whether they were perennials, sub-shrubs, shrubs or trees, all were planted in our garden at a size that is the ideal standard for a dry garden, in other words about 15–20cm (6–8in) high, so that the volume of the foliage is equal to the root volume. While we included arbutuses and cistuses, olive trees and euphorbias, oaks and sages, on looking at the garden from a distance one might have thought it was a single-layer planting of miniature plants.

Although the initial planting didn't look impressive because of the modest size of the shrubs and trees, the rapidity with which it developed was amazing. After only a few years the natural growth of the various types of plant had transformed the visual rhythm of the garden. From what seemed to be a uniform planted surface our garden had evolved into a space with a strong relief, created by alternating four types of zones:

• Open zones, consisting of several layers of low-growing plants: a flowering steppe planted with groundcovers and bulbs on a bed of gravel; a green terrace where a mixture of carpeting and pioneer plants, including *Sideritis cypria*, *Pallenis maritima* and *Stachys cretica*, grow in the cracks between the paving stones; a gravel garden consisting of cushion- and ball-shaped plants; path edges planted with perennials that tend to spread, such as valerians, asphodels, catananches, *Antirrhinum barrelieri*, *Stipa barbata* and euphorbias; and finally a wild unirrigated lawn which we scythe in late spring before it dries up partially or totally depending on how hot the summer is, consisting of a mixture of groundcover plants such as *Achillea crithmifolia*, *Achillea coarctata*, *Potentilla neumanniana* and *Trifolium fragiferum*, among which we welcome various weeds that spontaneously arrive seasonally to fill any empty spaces, such as pimpernels, black medick, hawksbeard and buck's horn plantain.

• Closed zones where large garrigue shrubs have in the space of about ten years made tall and dense plantings: in

Top: Planted as a small specimen in our garden, after 15 years this turpentine tree (*Pistacia terebinthus*) has grown into a magnificent specimen.

Bottom: An old *Pistacia lentiscus* in the Park of Saleccia. The garrigue garden becomes richer year by year without its evolution ever ending; the aim of the garden is not to achieve a hypothetical fixed planting but rather to observe its continual evolution, as fascinating in the early years as it is decades later.

these zones, which are scattered throughout the garden, the foliage of myrtles, lentisks, phillyreas, buckthorns, bupleurums, tree germanders, viburnums and bays blocks the view, thus dividing the garden into discrete sections with complementary atmospheres and allowing new views to open up as one walks through it.

• Zones that one can see through, consisting of a double layer of taller and low-growing plants: olive trees planted against a colour-coordinated background of groundcover plants with grey, grey-green and silver foliage, cypresses surrounded by a selection of plants that do well beneath large conifers, such as *Centaurea cineraria*, *C. argentea*, *Salvia interrupta* and *Helichrysum orientale*, pine trees emerging from an underplanting of plants that tolerate shade, the fall of pine needles and root competition, such as *Erica manipuliflora*, *Ampelodesmos mauritanicus*, *Prasium majus*, *Ptilostemon chamaepeuce* and a collection of hybrid cistuses that we chose for their ability to grow naturally in pine woods.

• And finally wilder zones, given structure by shrubs and trees which were very small when planted but which have now after some years become magnificent, such as *Cotinus*, *Vitex*, male and female nut-bearing pistachios (*Pistacia vera*), turpentine trees (*Pistacia terebinthus*), a hybrid evergreen pistachio (*Pistacia × saportae*), Judas trees (*Cercis siliquastrum*), almonds, holm oaks, flowering ash (*Fraxinus ornus*) and arbutuses, surrounded by different layers of lower-growing garrigue plants including ballotas, cistuses, germanders, *Dorycnium*, euphorbias, artemisias and sages, which self-seed abundantly between the tree medicks, lentisks, coronillas and viburnums.

The different zones of our garden have evolved at different speeds, with the various parts becoming visually linked as they developed. In the open zones, herbaceous plants and sub-shrubs quickly grew to present scenes that appear to be mature, and these have remained fairly stable thanks to the particular conditions that we deliberately created in order to maintain these first layers of low-growing plants in an equilibrium phase. In other parts of the garden the plantings have continued to evolve, with a very different appearance after five, ten, fifteen and twenty years. Having originally been planted with very small specimens which temporarily made the layout look uniform, the garden is now in the process of a two-fold evolution, both spatial and temporal, which is leading through successive stages to a strong visual differentiation between the zones. A garrigue garden is attractive when it is still young, then in the following years becomes even richer without its evolution ever reaching an end – for the aim of such a garden is not to achieve a hypothetical fixed result but rather to allow us to observe its continual evolution, remarkable in its early years and just as fascinating decades later.

WORKING WITH NATURE: MAINTAINING A GARRIGUE GARDEN

Page 150: The dynamic self-seeding of two pioneer species on a hillside: *Ferula communis* and *Hyparrhenia hirta*.

Clary sage (*Salvia sclarea*) self-seeds freely on a stony slope. In a garrigue garden, maintenance consists of working with nature as the garden is successively transformed.

In a traditional garden, maintenance is aimed at keeping the different garden scenes looking the way they were designed to be at the initial planning stage. Herbaceous borders need to be weeded, rose beds demand pruning and the lawn must be mown. In a garrigue garden, maintenance plays a different role. It is no longer a case of struggling to keep the garden looking as it did at the beginning but rather of going along with nature as the garden undergoes successive transformations. The creation of a garrigue garden is the starting point for several possible trajectories in its development. It is only the first stage of diverse gardens to come which will evolve over different time scales depending on the choice of species and planting techniques in the different zones of the garden. Their evolution may be guided by selective work on the part of the gardener, who may or may not choose to channel the

colonizing behaviour of various species. Although new plantings require regular supervision, the need for maintenance subsequently declines rapidly as the garden matures. What is more, far from being a constraint, maintenance then becomes a fascinating tutorial in gardening: in a garrigue garden, maintenance work is an apprenticeship that teaches us about the multiple ways in which plants interact with their environment.

• *Maintenance during the establishment phase*

The establishment phase after planting is the period that needs the most maintenance in a garrigue garden. To shorten this period, the gardener can help plants to become established quickly by applying optimal dry garden techniques. These techniques, which enable plants to become independent more rapidly, can be summed up in four points: decompaction and drainage of the soil in order that the young roots can grow down more easily in an oxygenated environment; choosing small specimens to help them become established; planting at the beginning of autumn to give the root system plenty of time to develop before summer sets in; making watering basins, which should remain in place until the end of the first summer, to allow deep watering which will draw roots downwards. Indeed, in a young garrigue garden the purpose of watering during the first year is not to encourage the growth of the parts of the plant above ground but rather to enable an extensive root system to develop, so that a balance between water loss from the foliage and water absorption by the roots is achieved as rapidly as possible. The first year is thus a latent period, during which a plant's development is not spectacular. During this stage, maintenance enabled us to observe our young plants closely while at the same time offering us our first surprises and rewards: a Corsican hellebore flowering at Christmas, the young shoots sprouting from the swollen rootstock of a caper at the end of winter, the unfurling of the hairy crosiers of clary sage in spring, or the emergence of a swallowtail butterfly in early summer, slowly unfolding its crumpled wings as it came out of the chrysalis still attached to a young fennel stem.

Maintenance during the first year consists of several activities: watering, carried out by filling the wide watering basins by hand about once every three weeks during the first summer if the planting has been done at the beginning of the previous autumn; occasional repair of the watering basins; and regular weeding beneath the plants to ensure that competition by weeds does not impede the young garrigue plants' root development. How much manual weeding will be required during the first year, when watering favours the germination of weeds, depends to a large extent on the nature of the soil. In land that was formerly agricultural the number of weeds may be significant. In gardens in residential areas that have been

Top: In a new planting, maintenance work gives the gardener the opportunity to observe the young plants closely which sometimes offers surprises and rewards: a swallowtail butterfly slowly unfolds its crumpled wings as it emerges from its chrysalis which is attached to the stalk of a young fennel.

Bottom: An agricultural landscape shaped by the use of chemical fertilizers and herbicides. In residential areas that were once farmland, the garden soil often has a long history of disturbances which favour a range of highly competitive nitrophile weeds: manual weeding during the first years can represent a significant amount of work.

created by the expansion of towns into what were once fields, for example, the soil often has a long history of disturbances which favour a specific range of highly competitive weeds. The regular application of fertilizers and chemical herbicides has often selectively fostered an array of nitrophile species that have developed a resistance to herbicides, such as horseweed, amaranth, white rocket,

Top: In the garrigue, the poor and stony soil favours garrigue plants but not the competing plants that appear in richer soils. *Santolina chamaecyparissus* in the calanques of Marseilles.

Bottom: Adding sand and gravel when preparing the soil improves drainage and impoverishes the soil, which reduces competition from weeds during the early years.

bindweed, fat hen and thistle: without realizing it, the new gardener has inherited a substantial seedbank in the soil. These seeds germinate easily in soil that has recently been dug before planting.

On the contrary, when the soil is naturally poor and stony, as in the garrigue, there are few weeds to compete with the young cistuses, euphorbias, phlomises and asphodels which start to grow in the garden between the stones. If the gardener has added sand or gravel to improve drainage during the preparation of the soil, he or she will benefit twice over, since the impoverishment of the soil coupled with good drainage also reduces the time that will need to be spent weeding during the first year. .

• *Maintenance during the growth phase*

From the second year onwards the garrigue garden starts to develop rapidly. As soon as the spread of the roots is sufficient the plants start to grow vigorously, often faster than one could have imagined. Pioneer garrigue plants, capable of colonizing open spaces in the tough conditions of their natural habitat, express their full growth potential in garden conditions, as long as the soil is properly drained and aerated. In just a few years the garrigue garden becomes unrecognizable. Cretan scabiouses form magnificent cushions, prostrate rosemaries cascade over walls, shrubby sages are covered in flowers, achilleas seed themselves between the stones, the dense mass of artemisias

Consisting as it does of drought-resistant wild plants, the garrigue garden quickly demonstrates how little maintenance is needed when irrigation is out of the picture. This scene includes *Salvia fruticosa, Euphorbia rigida, Lavandula × intermedia, Teucrium fruticans* and *Anthyllis cytisoides.*

gives structure to the space and the long inflorescences of giant stipas point skywards. During this growth phase very much less maintenance is required than in the initial establishment phase. Because it is composed of drought-resistant plants, the garrigue garden benefits from the reduction in maintenance that an unwatered garden makes possible. Once the plants are established it is no longer necessary to water them in summer, which means that the weeds encouraged in traditional gardens by irrigation during the hot time of year are eliminated. Thus in a dry garden weeding is eventually limited to seasonal interventions, since herbaceous plants germinate mainly in mild and damp conditions, in spring or autumn.

The level of maintenance that will be required in the various areas of the garden depends on the technical solutions adopted, as the gardener can intervene differently in different zones to guide the plant dynamics. The choice of species, the density of planting and mulching the surface of the soil between the plants all have a direct influence on maintenance during the growth years of garrigue plants. A low planting density allows one to walk freely among the garrigue plants. In this type of open garden the use of an inorganic mulch limits the germination of weeds, making maintenance requirements minimal. The inorganic mulch is generally added one year after planting, when the watering basins – now no longer necessary – have been levelled. In our experience, the optimal thickness of a gravel mulch is 6–8cm (15–20in), with the grain size being about 5–20mm (¼–¾in). This type of washed gravel, which does not contain fine particles, has a twofold influence on plant dynamics. On the one hand it limits the germination of the herbaceous weeds that one finds on agricultural land, while on the other it favours the germination of garrigue plants which self-seed naturally in stony soil. Making a gravel garden is thus one of the best solutions for reducing weeding while at the same time encouraging the self-seeding of garrigue plants in the garden.

As an alternative to an inorganic mulch, another way of managing the soil surface when the planting density is low is to scythe it occasionally. The herbaceous plants that appear in the garden spontaneously will then form seasonal green corridors between the young garrigue shrubs. When the soil has become stabilized after a few years, this range of herbaceous plants will be very different from the pioneer weeds that exploited the disturbance of

the soil during planting. At this stage, spontaneous herbaceous plants can live in harmony with garrigue shrubs, often giving rise to positive interactions between the species: some keep the soil well oxygenated thanks to their taproots which break up the soil at a depth; others live in symbiosis with a range of bacteria and beneficial fungi, contributing to the biological life of the soil; and finally many herbaceous plants, although sometimes of modest appearance, bear flowers that are rich in nectar and pollen, which are major sources of food for a whole range of insects that are useful in a garrigue garden. In our own garden, slender sowthistle (*Sonchus tenerrimus*), salsify (*Tragopogon porrifolius*), hawksbeard (*Crepis sancta*), black medick (*Medicago lupulina*), smooth golden fleece (*Urospermum dalechampii*), common melilot (*Melilotus officinalis*), mallows (*Malva sylvestris*) and wild chicory (*Cichorium intybus*) open their brightly coloured flowers in the free spaces, increasing the populations of pollinators and attracting numerous species of hoverflies, lacewings and ladybirds which play an essential role in the food chain of the garden.

In some zones of the garden one can choose to plant more densely, so that in a few years the evergreen shrubs will completely cover the soil. In this type of closed landscape weeding is reduced to almost zero, since the dense vegetation no longer leaves any room for weeds. If the plants are tall enough when mature – more than 40–50cm (16–20in) – the creation of a closed bed consisting of evergreen plants is also the most effective way of suppressing perennial weeds that spread by means of rhizomes, such as dandelions or bindweed, which under certain conditions can show a strong colonizing tendency in an open environment. When one is making a closed planting, one can plan the rate at which the ground will be covered by choosing a planting density suitable for different species. Some garrigue plants will attain a height not far from their mature height in only a few years in favourable conditions where the soil has been deeply decompacted. *Anthyllis cytisoides*, arborescent artemisias (*Artemisia*

arborescens and *A. thuscula*), Sicilian euphorbia (*Euphorbia ceratocarpa*), shrubby sage (*Salvia fruticosa*) and the hybrid lavender *Lavandula × allardii*, for example, are plants that can grow very fast: a planting density of one plant per 1sq m (11sq ft) will provide total cover in just three or four years. However, a greater planting density can be chosen instead if covering the ground is a matter of urgency for the gardener who wishes to reduce maintenance as quickly as possible.

Some species that grow rapidly from their first years can become very much larger when they are mature. Most prostrate rosemaries, for example, are capable of covering an area of more than 4sq m (43sq ft) after ten years or so. The dense mass of a tree germander (*Teucrium fruticans*) can take up about 1sq m (11sq ft) of soil after a few years, then become an imposing irregular ball with a diameter of almost 3m (10ft) after 15–20 years. We have seen an old specimen of shrubby scabious (*Lomelosia minoana*) growing on the slope above a road in the mountains of Crete, where it formed a monumental groundcover spreading over an area of more than 15sq m (161sq ft). The usual planting density chosen for these plants that in time can become very large is still one plant per 1sq m (11sq ft) when one is aiming to achieve rapid cover, even though this is obviously too great a density in relation to the plant's long-term potential growth. When in the future this planting

Euphorbia ceratocarpa is a fast-growing plant which can cover more than 1sq m (11sq ft) in three or four years.

Following pages: In areas of the garden planted at a high density, the need for weeding can be reduced to almost zero after a few years since the plants no longer leave any space for weeds. This scene mingles *Santolina neapolitana* 'Edward Bowles', *Stachys cretica*, *Coronilla minima*, *Origanum syriacum*, *Catananche caerulea* 'Tizi N'Test', *Erica manipuliflora* subsp. *anthura*, *Euphorbia ceratocarpa*, *Anthyllis cytisoides* and *Teucrium fruticans* 'Agadir'.

Top: Some fast-growing plants can cover an area of about 1sq m (11sq ft) after a few years, then in the medium term become even larger. This planting between cypress trees consists of prostrate rosemaries (*Rosmarinus officinalis* 'Boule'), tree germanders (*Teucrium fruticans*), phillyreas (*Phillyrea angustifolia*), broom (*Spartium junceum*), bay (*Laurus nobilis*) and atriplex (*Atriplex halimus*). Richard and Jola Gillespie's garden designed by Jennifer Gay and Piers Goldson on the island of Andros.

Bottom: When they find the poor, well-drained soil conditions that allow them to live for a long time, some garrigue plants can become very large indeed. This shrubby scabious (*Lomelosia minoana*), bearing thousands of flowers, grows on the south-facing side of Mount Dikti in Crete, where it forms a monumental groundcover covering an area of more than 15sq m (161sq ft).

has evolved, some plants will be smothered by the stronger vegetation of species that progressively come to be dominant.

Other garrigue plants, of interest for their robustness, long lifespan and the quality of their foliage or flowers, are by contrast slow-growing at first. The magnificent lime-tolerant heathers (*Erica multiflora* and *E. manipuliflora*), woody gromwell (*Lithodora fruticosa*), olive-leaved convolvulus (*Convolvulus oleifolius*), Balearic St John's wort (*Hypericum balearicum*) and the cushions of thorny euphorbia (*Euphorbia spinosa*) grow slowly, taking years to show the potential volume of their vegetation. When one is trying to create a closed planting, these species can be planted in groups of three or five at a greater density to occupy islands with small areas. Alternatively, they can be reserved for open types of garden, as for example gravel gardens, where they will have all the time they need to develop without competition thanks to the inorganic mulch.

One way of covering the soil and limiting maintenance while waiting for the evergreen structure of garrigue shrubs to grow large enough is to fill empty spaces temporarily with plants that have a shorter life span. One is thus working on a double time scale, the first seeking to occupy the space in the short term, the second looking ahead to the plants which will take over after a few years. This type of management, right from the planning stage, envisages a complete transformation of the garden's appearance over its early years: the structural plants will become dominant after four or five years, while the pioneer plants will occupy the space at the beginning. These fast-growing pioneer plants include semi-woody as well as perennial species.

Dorycnium hirsutum, for example, is a sub-shrub that establishes quickly in decompacted soil, is covered in white flowers from its first year, and forms a substantial spreading cushion in less than two years. Its short life span of about three to four years is compensated for by the fact that it produces a large number of seeds and self-sows easily on gravelly ground. *Salvia desoleana* is another sub-shrub that rapidly forms a thick mass of large, velvety, scented leaves topped by strong inflorescences bearing bi-coloured flowers, light blue and pale yellow.

Herbaceous pioneer plants are equally abundant in a garrigue garden. Various species of valerian all grow fast and flower from their first year, for example common valerian (*Centranthus ruber*), with flowers that may be red, pink, white or mauve, Lecoq's valerian (*C. lecoqii*), which is happy in stony soils, and the glaucous-leaved *C. longiflorus*, which is exceptionally drought-resistant. The inflorescences of Cupid's dart (*Catananche caerulea*) form a light mass that is perfect for temporarily filling the gaps between shrubs and it self-seeds every year into bare spaces. Some herbaceous species spread by rhizomes, such as the handsome silver-leaved *Achillea coarctata* and *A. crithmifolia* – the latter is capable of spreading with astonishing speed in soil that has recently been worked. In some areas of our garden we have allowed young plantings to be invaded by a flowery and aromatic mixture of pitch trefoil (*Bituminaria bituminosa*), calaminth (*Calamintha nepeta*), clary sage (*Salvia sclarea*) and fennel, creating a scene in which these wild-looking herbaceous plants occupy the space for a few years before being progressively driven out by the growth of the evergreen garrigue shrubs.

Top: The inflorescences of Cupid's dart (*Catananche caerulea*) form a light mass, perfect for temporarily occupying the gaps in beds that are in the course of being planted.

Bottom: On the Pingus wine estate in Spain a temporary herbaceous layer, dominated by *Achillea nobilis* and *Anthemis tinctoria*, was sown between the plants that will provide the future structure of the garden. Designed by the landscape architect Tom Stuart-Smith. Photograph by Peter Sissek.

The landscape architect Tom Stuart-Smith designed a striking arrangement on the vine-growing estate of Pingus, near Valladolid. A vast garrigue garden was planted on the slopes surrounding the buildings on the property. It was conceived of as a space consisting of several layers of interlinked and evolving vegetation, intended to be displayed in successive stages. A framework of saplings was first of all put in on the hillside: Portuguese oaks (*Quercus faginea*), holm oaks (*Quercus ilex*), field maples (*Acer campestre*), almond trees and Judas trees (*Cercis siliquastrum*). Between these future trees, the main layer consists of a collection of garrigue shrubs – cistuses, santolinas, thymes, sages, helichrysums and euphorbias – planted at a relatively low density, given the size of the garden. To cover the soil while the shrubs were growing, a second layer of herbaceous plants, a mixture of annuals, biennials and perennials, was added by sowing their seeds directly in situ. These plants, chosen for their successive flowering seasons and attractiveness to insects, include viper's bugloss (*Echium vulgare*), lady's bedstraw (*Galium verum*), ox-eye chamomile (*Cota tinctoria*), yarrow (*Achillea nobilis*), Cupid's dart (*Catananche caerulea*), oregano (*Origanum vulgare*), silky-spike melic (*Melica ciliata*) and lamb's ears (*Stachys byzantina*). The seeds were sown in autumn, just after the shrubs were planted, in order for them to germinate rapidly in the mild autumnal conditions. By the following spring a spectacular mass of flowers had taken over the space, making the garrigue plants almost invisible for a few months.

The dynamics of evolution in a garrigue garden: the herbaceous layer, which includes *Achillea nobilis* and *Centranthus ruber*, will gradually be replaced by evergreen shrubs such as *Lomelosia minoana*, *Artemisia arborescens*, *Phlomis viscosa*, *Santolina rosmarinifolia* and *Salvia pomifera*, which colonize the ground beneath a Mount Etna broom (*Genista aetnensis*) and chaste tree (*Vitex agnus-castus*).

In this type of management, maintenance during the first years consists of a late scything, in about July, in order to reduce competition for water in summer while at the same time making the most of the abundant seeds produced by herbaceous plants as they seek to self-seed wherever they find a place between the garrigue shrubs. During the first summer, additional maintenance is needed to reshape and clear the watering basins temporarily overrun by the herbaceous plants. In time, these herbaceous plants will give way to the evergreen shrubs, which themselves will one day be dominated by the future tree structure of the garden.

• The first phase of maturity: weeds and wild plants

The architecture of evergreen plants is revealed fully after four or five years, when the garrigue garden reaches its first phase of maturity. At this stage the cushions of densely foliaged sub-shrubs, such as Sardinian santolinas (*Santolina insularis*), the hybrid lavender *Lavandula × chaytorae* and Mount Hymettus scabiouses (*Lomelosia hymettia*) are already giving rhythm to the space. Intermediate shrubs such as *Salvia pomifera, Rosmarinus × mendizabalii* and *Cistus halimifolius* have attained their full size, and taller shrubs such as shrubby hare's-ear (*Bupleurum fruticosum*) and bridal veil broom (*Retama monosperma*) are emerging from the tight layer of cistuses and sages. The glossy foliage of sclerophyllous plants – phillyreas, lentisks, myrtles and arbutuses – with markedly slower initial growth is beginning to take their place, making a dark counterpoint to the grey or golden plantings of artemisias and phlomises.

When the garden has reached this first phase of maturity very little maintenance is required. The management of weeds is reduced to practically nothing both in the zones of closed planting, now entirely covered by the dense vegetation of garrigue shrubs, and in the open zones, designed as gravel gardens, where the spaces between the plants are covered with an inorganic mulch. The gravel garden can also be home to less competitive plants: bulb species such as crocuses, *Sternbergia*, anemones, wild tulips,

alliums and narcissi, as well as rock plants with attractive flowers, such as the white daisies of *Rhodanthemum hosmariense*, the purple flowers of *Teucrium cossonii*, the silky pink buds of *Convolvulus oleifolius*, the lemon-yellow spikes of *Sideritis syriaca*, the metallic blue bracts of *Eryngium amethystinum* and the delicately fringed flowers of *Dianthus broteri*. However, unlike the mixed borders of English gardens which are meant to remain 'clean' and weed-free throughout the year, a garrigue garden on gravel can perfectly well host the spontaneous plants that turn

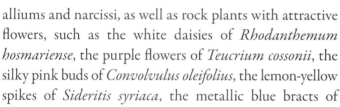

Plants once considered weeds are today often valued for the part they play in urban areas.

up by themselves amid the garrigue species, so that the inflorescences of maritime scabiouses (*Scabiosa atropurpurea*) or of love-in-a-mist (*Nigella damascena*) mingle with the flowers of wild plants deliberately introduced into the garden, such as bloody cranesbill (*Geranium sanguineum*), *Salvia verbenaca*, *Nepeta tuberosa* and *Stachys germanica*, without one really being able to distinguish the wild plants from the cultivated. The distinction between wild plants considered to be weeds and wild plants considered as ornamental species thus becomes so fuzzy that the gardener can apply other criteria for maintenance, as for example the plant's intrinsic qualities: its drought resistance, its prolonged flowering in tough conditions, or its attractiveness to beneficial insects. Maintenance then becomes a question of controlling the species that colonize most profusely just to make room for all the plants in the garden. This change in attitude, which partially blurs the distinction between weeds and 'good' plants, is what makes it possible to limit maintenance in a garrigue garden.

This new way of managing weeds can be applied both in private gardens and in public green spaces. Some plants that were previously considered weeds are today highlighted by nature societies. An example is the educational project 'Sauvages de ma rue' (Wild Plants in My Street), started in France by the Muséum national d'histoire naturelle, which encourages people to look at weeds in urban areas from a functional point of view, considering for instance how wild plants in cities can contribute to reducing air pollution, lowering the temperature via heat islands and maintaining pollinator populations. Whereas once a simple milkweed growing in a crack in a gutter might have been seen as indicating a lack of maintenance in public green spaces or the road network, this type of educational project invites people on the contrary to admire the way plants, no matter how humble, such as milkweed, plantain, hawkweed, pellitory or celandine, are willing to grow and flower in environments as hostile as a crack in concrete or at the base of a wall. This new way of looking at things can also be applied to the wild plants that establish themselves on the edge of plantings. In a public green space planned as a garrigue garden, consisting of a mixture of woody and herbaceous wild plants, the complementary wild plants arising from seeds borne by wind, water, birds or on the soles of visitors' shoes can blend in perfectly with the original planting.

• *The dynamics of self-seeding*

Once the garden has entered its first phase of maturity, the self-seeding of garrigue plants often becomes significant. The gardener can guide the dynamics of this seeding to change the arrangement of plants in the beds progressively, according to his or her inspiration and depending on the framework of the initial planting. The maintenance tasks of a traditional garden – mowing, watering, applying fertilizers or herbicides – are replaced by work of a different kind: the maintenance now first consists of a period of observation to decide which of the self-sown seedlings are to be preserved and which removed. This selection process can take place in autumn, winter or spring – it doesn't matter which – when the damp soil makes the seedlings easier to pull up.

A great many species self-seed naturally in a garrigue garden, each of them having its own strategy to move away from the mother plant, seeking dispersal to the most propitious spot for germination. The seeds of *Euphorbia rigida*, *E. corallioides*, *E. characias* subsp. *wulfenii* and *E. ceratocarpa*, scattered all around when their ripe seed capsules explode, are then further dispersed by the indefatigable activity of ants. Sometimes dropped as the ants scurry backwards and forwards to their nests, these seeds germinate all over the garden. The inflorescences of candelabra sage (*Salvia candelabrum*), clary sage (*S. sclarea*) and sticky sage (*S. viscosa*) become stiff canes full of seeds when they dry out after flowering. When the wind shakes them these tall inflorescences act as catapults, expelling their seeds far and wide, extending the areas of soil over which the fine rain of black seeds falls. The light seeds of valerian (*Centranthus ruber*), of *Ptilostemon chamaepeuce* and of the Cretan *Staehelina petiolata* are equipped with bundles of silky hairs that enable them to travel randomly through the garden, carried by the wind. Those of the yellow germander (*Teucrium flavum*) and fringed rue (*Ruta chalepensis*), which are pioneer species able to self-seed in the poorest soils, simply fall beneath the mother plant and germinate near it, sometimes remarkably generously. In this way the yellow germander, starting from a single isolated plant, can progressively form a large planting, extending by a few dozen centimetres every year, from the hundreds of seedlings that germinate in autumn in the space around the mother plant.

The seeds of *Stipa barbata* are gifted with an ability to creep, allowing them to advance over the soil by themselves. It is the wind that carries the seeds away from the mother plant in the first place, thanks to the long silky beard that is a prolongation of the seed. This beard subsequently coils up like a fine spring, the coiling modified during the daytime by variations in the atmospheric humidity. The seeds use this coiling and uncoiling of the spiral to advance over the ground until they find a gap between the stones into which they can screw themselves down deeply. The fine seeds of the horned poppy (*Glaucium flavum*) have photosensitive pigments which enable the plant to choose an environment

Spontaneous seeding of euphorbia (*Euphorbia characias* subsp. *wulfenii*): in our garden a real explosion of seedlings appears in autumn in empty spaces, mostly on inorganic mulches and gravel paths.

where it can develop with weak competition. The seeds germinate only in porous soils, generally consisting of sand or gravel, where they can bury themselves by slipping between the large particles in the soil, washed down by water as it infiltrates the ground. They can then germinate in complete darkness, their germination being prevented by light. In our garden we fairly often see young horned poppies appearing round new plantings when we have prepared the soil by adding sand and gravel to improve drainage.

The countless strategies plants use to disperse their seeds, such a source of wonder to the gardener, are an expression of the incredible power of garrigue plants to self-sow. In our garden a whole crop of seedlings appears every autumn in any empty spaces. Aspic lavenders (*Lavandula latifolia*), both green- and grey-leaved phlomises (*Phlomis longifolia*, *P. fruticosa*), *Dorycnium pentaphyllum*, shrubby sages (*Salvia fruticosa*), Cretan cistuses, santolinas, ballotas and Algerian irises (*Iris unguicularis*) self-seed every year in a mixture of tiny plantlets that appear mainly on top of the inorganic mulch or along the gravel paths. In a garrigue garden this self-sowing dynamic enables the gardener to establish interesting bridges between the planning stage and the maintenance of the garden. Simply by choosing which young plants to keep and which to discard, the gardener progressively redesigns his or her plantings. Thus it is this kind of maintenance that fine-tunes

Top: *Stipa barbata* advances every year in our garden: its seeds have an ability to creep which enables them to move over the soil completely unaided.

Bottom: The gardener can progressively redesign plantings by selecting the seedlings he or she will preserve: in this way maintenance becomes more of a planning project which continues year after year.

the original plan, for plants that move around often after a few years end up finding positions which suit them much better than if their places had been fixed at the time of the initial design.

• *Guiding the garden's evolution*

As a further step after choosing which self-sown seedlings to keep, the garden's evolution can be guided by selective clipping of the plants. In the garrigue garden, this serves complementary purposes. For some species, cutting back after flowering showcases the forms and textures of the foliage, since the faded inflorescences may mask a plant's architecture. Some garrigue sub-shrubs, such as cistuses, helichrysums, French lavenders and santolinas, also live longer if they are clipped occasionally. In the course of their evolution in the wild these plants have become adapted to being browsed by sheep and goats. A light clipping, which mimics the effects of grazing, prevents the base of the plant from becoming bare and enables it to have a longer life span.

However, one may choose not to cut back other species after flowering in order to enjoy for longer the diversity of the silhouettes and colours of their dry inflorescences. The silver mist of Cupid's darts (*Catananche caerulea*), the flowers of *Helichrysum orientale* that turn golden as they age, the tight mass of the magnificent dry inflorescences of shrubby scabiouses, the tiered brown inflorescences that surmount the green, russet, white or golden leaves of phlomis species (*Phlomis viscosa*, *P. bourgaei*, *P. nissolii*, *P. × termessi*), the huge dry umbels of the giant fennel that rise above the plantings, the delicate little spikes of *Lygeum spartum* shaped like a bird's beak, the inflorescences always in motion of *Piptatherum miliaceum*, the light, fountain-like silhouette of *Stipa calamagrostis* and the long, supple stems of *Ampelodesmos* bending under the weight of their golden seeds – all these draw the eye in the garrigue garden when not much is in flower. If some plants are worth clipping in summer to reveal the beauty of their foliage, other plants by contrast can be left for a long time in the natural, free habit that shows them off best. Selective clipping thus contributes to the rhythm of the landscape, with an alternation between clipped and non-clipped plants adding to the visual attraction of a garrigue garden in summer.

Clipping also makes it possible to guide the garden's evolution by controlling the self-seeding of garrigue plants. Depending on the date at which it is carried out, clipping can prevent a pioneer plant from multiplying. If you cut back a yellow germander (*Teucrium flavum*) in autumn or winter, for example, the seeds will have had all the time they need to scatter during the summer. But if instead you cut the yellow germander back in early summer, just when its flowering is almost over, you get rid of the seeds before they are ripe, preventing all possibility of self-seeding. Selective trimming can thus be carried out in the garden as flowering progresses in order to define the spaces at any given moment in the garden's evolution where one does or does not want self-seeding. For a few years now, for example, seedlings of the lovely candelabra sage (*Salvia candelabrum*) have become so abundant in one part of our garden that we have decided to put a brake on their spreading for the time being by cutting off their faded inflorescences in June, before the seeds are dispersed. It is easy to limit in this way the colonizing tendencies of plants whose pioneering character in the garden can sometimes be impressive, not only in herbaceous plants such as clary sage, fennel, *Stipa gigantea* and *Euphorbia ceratocarpa*, but also in shrubby species such as coronilla (*Coronilla valentina* subsp. *glauca*), phlomis, viburnums and bupleurum.

To guide the evolution of the garden, other technical solutions can complement controlling young seedlings and clipping selectively. In the garrigue, the landscape sometimes enters temporary equilibrium phases as a result of different pressures which have a local effect on the vegetation. The spatial distribution of the floral mass in the

1. Most garrigue sub-shrubs such as lavenders, rosemaries, helichrysums and santolinas live longer if they are clipped occasionally. The garden at Rou, Corfu, designed by Jennifer Gay and Piers Goldson.

2. *Phlomis grandiflora* in summer: if some plants are worth clipping during the summer to reveal the beauty of their foliage, others can be left longer to grow freely and show off their dry inflorescences.

3. The golden inflorescences of *Ampelodesmos mauritanicus* in summer at the tip of Punta Campanella, south of Naples.

4. As they fade, the flowers of Cupid's dart (*Catananche caerulea*) give way to silvery seedheads which catch the light and remain ornamental all summer long.

landscape is modified, but without any major alteration of the structure of the vegetation. It is possible to create scenes inspired by these landscapes in equilibrium within the garden too. In areas subject to regular foot traffic, the relationship between mineral surfaces and plant surfaces can be one of the ways to control the evolution of the vegetation. Contemporary landscape designers are exploring new models for green terraces inspired by the remains of old threshing floors that have been partially colonized by garrigue plants. In this type of management, it is stone that is dominant, with plants simply slipping into the cracks between the paving. By choosing plants able to grow in these cracks, one can create open spaces that require minimal maintenance since the stone surface blocks the evolution of the landscape.

In the gardens of the Fort Saint-Jean in Marseilles, for instance, the landscape architects of the APS Agency designed a stunning green terrace in the northern part of the parade ground, visited daily by hundreds of people. It evokes a threshing floor overrun by vegetation but here the species planted in the cracks thrive between the paving stones, making remarkable miniature gardens of striking beauty. Illustrating the new interest in the balance between stone and plants in contemporary gardens, in May 2017 the landscape designers James and Helen Basson won the award for Best Show Garden at the famous Chelsea Flower Show for their garden inspired by an abandoned quarry in Malta, with a subtle dialogue between the wild garrigue plants that self-seed freely between stones and the geometric design of the blocks of limestone left lying in the quarry.

In our garden, we have tested the behaviour of many plants that are able to colonize a green terrace consisting of large paving stones laid on a bed of sand. When we made the terrace, we started out with a mixture of three complementary types of plant: carpeting groundcovers capable of multiplying between the paving stones, as for example achilleas, thymes, potentillas, sedums, *Artemisia*

5. The dry inflorescences of asphodels emerging from luminous cushions of helichrysums contribute to the beauty of this landscape in the mountains of Corsica, near Corte.

Left: Selective clipping makes it possible to guide the garden's evolution by controlling garrigue plants' self-seeding potential.

pedemontana and dwarf irises; cushion-shaped plants spilling irregularly over the paving, such as stachys, germanders, centaureas, tansies and oreganos; and more slender plants with silhouettes that stand out, such as asphodels, asphodelines and snapdragons, which self-seed to fill any available space.

In zones with a lot of foot traffic, it is the fact of being walked on that controls the ebullience of the vegetation, but many garrigue plants with a carpeting habit show an unexpected tolerance to this. In zones with less foot traffic, the evolution of the scene is naturally limited by the narrowness of the cracks in which the plants grow. The maintenance of this kind of green terrace is minimal – simply removing faded inflorescences is enough to keep the terrace visually attractive throughout the year.

The success of this type of planting is due in part to the fact that the plants' roots benefit from the humidity that remains underneath the paving for a long time, since the stone slabs prevent loss of water by direct evaporation from the soil. The plant palette that can be used for a green terrace in a garrigue garden is thus very wide. It is based on the numerous species that grow in a stony environment, squeezed between the stones of the garrigue or able to thrive in narrow fissures in the rocks.

In public green spaces there are other techniques that can be employed to reduce maintenance while at the same time limiting plant dynamics. Hard surfaces in urban areas are broken up by a multitude of open surfaces, the maintenance of which has become a major question for local authorities. The spaces around the bases of street

Top: Inspired by the spontaneous vegetation growing between the blocks of limestone in an old quarry on the island of Malta, this garden, designed by James and Helen Basson, won the award for Best Show Garden at the famous Chelsea Flower Show in May 2017.

Bottom: The roots of plants benefit from the moisture trapped beneath the paving stones of a green terrace.

trees or the narrow spaces between graves in a cemetery are examples of surfaces where maintenance may be difficult to manage, and new regulations banning the use of pesticides have led park services to seek alternatives to chemical herbicides. In some cases all that is needed is a different mindset, allowing weeding to be replaced by an acceptance of the 'weeds' that make up the spontaneous flora. In other situations, groundcovers which require less maintenance can be used to cover open spaces, the ideal groundcover in this context being a plant that can tolerate a certain amount of foot traffic, is long-lived and does not need either irrigation or weeding. Numerous drought-resistant garrigue plants are perfect candidates for this type of management, thanks to their allelopathic properties (see p.175) that limit the germination of competing species naturally.

Top left: Groundcovers reduce the need for maintenance in urban green spaces. This view of the park at the Stavros Niarchos Cultural Centre in Athens includes *Teucrium luteum*, *Tanacetum densum*, *Helichrysum italicum* subsp. *microphyllum*, *Achillea umbellata* and *Euphorbia spinosa*.

Top right: An allelopathic landscape in Corsica dominated by *Rosmarinus officinalis*, *Stachys glutinosa*, *Teucrium capitatum* and *Helichrysum italicum*.

Bottom: An allelopathic landscape in Dalmatia: occupying all the space, *Brachypodium retusum* and *Salvia officinalis* temporarily block the evolutionary trajectory of the landscape after a fire thanks to their allelopathic properties which limit the germination of competing species.

Allelopathy: a gift from the garrigue

Jerusalem sage (*Phlomis fruticosa*) has two distinct types of leaves with a function that changes according to the season. In spring, large leaves are held horizontally, to catch the sun's rays and optimize photosynthesis. At the beginning of summer these leaves fall and are replaced by much smaller leaves, which by contrast are held vertically to reduce their exposure to the sun. During summer the foliage of the Jerusalem sage is thus markedly less dense: the light passing through the leaves can reach the soil, which could make conditions favourable for seeds of other species to germinate beneath the plant with the first rains of autumn. To limit possible competition from plants germinating beneath it, over its evolution *Phlomis* has developed an effective strategy: as the leaves that fell to the ground in early summer decompose, they release chemical compounds that inhibit the germination of competing species.

This phenomenon, called allelopathy (from the Greek *allelon*, reciprocal, and *pathos*, suffering), is found in different forms in many garrigue plants. Allelopathy comprises complex mechanisms which in the wild influence the chemical interactions between plants, either positive or negative. In the context of the garden, when we refer to allelopathy we are usually thinking of negative allelopathy which enables a plant to defend itself against competition from other species. The ways in which allelopathy could be used in horticulture have so far been little studied but, drawing inspiration from the interactions of plants in their natural habitat, we might envisage allelopathy in the garden as a new way of limiting the germination of weeds and reducing the task of weeding.

Allelopathic species from garrigues include many of the plants currently used in gardens, such as cistuses, santolinas, phlomises, rosemaries, lavenders, thymes, savories, sages and artemisias. Within these genera, the diversity is so great that it enables us to conceive of very varied scenes. Cistuses, for example, include carpeting plants (*Cistus* × *lenis*, *C. crispus*), cushion-shaped plants (*C.* × *pauranthus*), plants that make large, regular balls (*C.* × *pagei*, *C.* × *florentinus*) and shrubs with an upright habit whose leaves can be dark green (*C.* × *cyprius*), grey (*C.* × *picardianus*) or silver (*C.* × *tardiflorens*). The diversity of phlomis species is notable, with flowers of pale yellow (*Phlomis russeliana*), orange-yellow (*P. monocephala*), white (*P. purpurea* 'Alba'), pink (*P. italica*) and bicoloured pink and white (*Phlomis bovei* subsp. *maroccana*).

The many varieties of rosemary (*Rosmarinus*), prostrate or upright, range from pure white through pastel pink and sky blue to an intense mauvey-blue. Thyme species have flowering periods that are spread over several months, with common thyme (*Thymus vulgaris*) flowering from April, Thracian thyme (*T. thracicus*) and camphor-scented thyme (*T. camphoratus*) in May, savory-leaved thyme (*T. saturejoides*) and mastic thyme (*T. mastichina*) in June, and finally the magnificent Cretan thyme (*Thymbra capita* syn. *Coridothymus capitatus*) flowering throughout July. The foliage of santolinas may be grey (*Santolina chamaecyparissus*), grey-green (*S. villosa*), silver (*S. magonica*), bright green (*S. rosmarinifolia*) or blueish in hue (*S. rosmarinifolia* 'Caerulea').

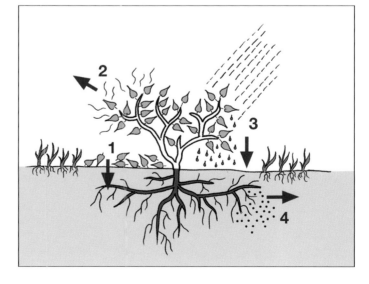

Allelopathic compounds may be diffused in different and sometimes complementary ways within the same species.
1. By the decomposition of the leaf litter that accumulates below the plant, as in cistuses and phlomises.
2. By depositing on the soil a fine allelopathic dew coming from the plant's essential oils, as in myrtles and some sages.
3. By rain washing down the compounds from the essential oil glands contained in the leaves, as in rue.
4. By secretions given off by the roots, as in common thyme and mouse-ear hawkweed.

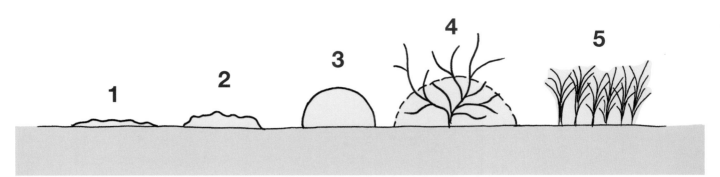

The allelopathic plants that can be used to cover the ground in gardens and green spaces include plants with very different forms and heights.
1. Carpeting plants a few centimetres high, such as *Thymus praecox*, *Pilosella officinarum* (formerly known as *Hieracium pilosella*) and *Achillea crithmifolia*.
2. Taller carpeting plants, about 10–15 centimetres (4–6in), such as *Psephellus bellus*, *Achillea umbellata* and *Tanacetum densum*.
3. Cushion- and ball-shaped plants, such as Cretan thyme (*Thymbra capitata*), *Phlomis × cytherea* and compact or prostrate rosemaries.
4. Sclerophyllous plants that tolerate regular clipping and form large rounded masses, such as lentisk (*Pistacia lentiscus*) and myrtle (*Myrtus communis*).
5. Plants with an upright silhouette emerging from lower plants, such as *Hyparrhenia hirta*.

Thymus praecox

Psephellus bellus

Phlomis × cytherea

Pistacia lentiscus

Hyparrhenia hirta

The diversity of sages found around the Mediterranean is quite simply amazing: their aromatic foliage and often spectacular flowering make them the plants of choice in the garden. They include, for example, little-known species such as the Cyprus sage (*Salvia dominica*), which has white, intensely aromatic foliage, the tomentose sage (*S. tomentosa*), which comes from the mountains of Lebanon and Syria, and the Sierra Nevada sage (*S. lavandulifolia* subsp. *vellerea*), which forms a remarkable groundcover in poor and stony soils.

All the above species show that at the very heart of the garrigue there is a whole range of plants with allelopathic properties that are useful in the garden. The garrigue garden thus lends itself particularly well to the creation of a series of allelopathic scenes which require almost no weeding once the plants are established.

The efficacy of a plant's allelopathy increases as it ages: it needs to have a certain volume of foliage and roots for its allelopathic properties to function. In areas of the garden planned to include allelopathic plants, one needs to consider how to manage weeds during the first few years after planting: according to the species, allelopathy generally starts to act only after three or four years into the plant's development. Depending on the planting density and the rate of growth of the species chosen, it is often necessary to apply a mulch in order to keep maintenance down while waiting for the plants' allelopathic properties to take effect. In time, this allelopathy will combine with the dual action of the mulch and the covering of the soil by the thick evergreen foliage. To help plan allelopathic plantings, useful information on rate of growth, planting density and the thickness of the foliage of the various allelopathic species that can be used in the garrigue garden is given in our database (see p.276).

• *Allelopathic gardens*

The town of Auch has planted allelopathic groundcover beneath the famous avenue of plane trees along Boulevard Sadi-Carnot, which runs through the heart of the town. A range of drought-tolerant plants, *Euphorbia myrsinites*, *Tanacetum densum*, *Achillea crithmifolia* and *Pilosella officinarum*, was chosen to cope with the constraints of this environment: the shallow soil and the competition from tree roots reproduce here, in a non-Mediterranean climate, conditions that are similar to those of the garrigue. After the soil was dug to a depth of 15–20cm (6–8in) and a mixture of earth and gravel was added to improve drainage, a gravel mulch was spread directly on the soil before planting. The groundcovers were then planted through the gravel, which made managing weeds easier

during the plants' first years, since their allelopathic action became effective only when they had had time to develop sufficiently. The planting has been a success in the view of both the gardeners and the public: the carpets of vegetation gained after a few years mean that very little maintenance is required today, freeing the gardeners from the meticulous task of weeding, while information is provided to the public on this planting which makes herbicides redundant.

Top: In the town of Auch, allelopathic groundcovers have been planted beneath plane trees to reduce the task of weeding. *Achillea crithmifolia* covers the ground completely and requires little maintenance.

Bottom: In the spaces between the graves in the new cemetery of Saint-Marc-de-Jaumegarde near Aix-en-Provence, allelopathic groundcovers were planted to reduce maintenance. The plant palette used includes *Psephellus bellus, Achillea umbellata, Tanacetum densum, Thymus camphoratus* and *Teucrium marum*. Designed by Sandrine Lefèvre and the architectural practice Mossé and Gimmig.

In the spaces between the graves in the new cemetery of Saint-Marc-de-Jaumegarde near Aix-en-Provence, a similar technique has been used. For this allelopathic groundcover planted on gravel a variety of garrigue plants was chosen, including *Salvia lavandulifolia* subsp. *blancoana*, *Teucrium marum*, beloved of cats, *Achillea umbellata*, thyme (*Thymus munbyanus* subsp. *ciliatus*), the carpeting *Psephellus bellus* (syn. *Centaurea bella*) and *Brachypodium retusum*. Amid the groundcover, other plants were introduced to add the seasonal interest of their leaves and flowers. These included hollow-stemmed asphodel (*Asphodelus fistulosus*), *Artemisia pedemontana*, *Euphorbia cyparissias* and the long-flowering and decorative *Limonium pruinosum* and *Goniolimon speciosum*. Designed by the landscape architect Sandrine Lefèvre and the architecture practice Mossé and Gimmig, this striking management of the space, in which the repetitive rhythm of the gravestones combines with the supple movement of the plants, makes a scene of great beauty in the unusual context of a cemetery while requiring little maintenance, no irrigation and no herbicides.

Whether in public green spaces or in private gardens, garrigue plants with allelopathic properties can be used in many ways to reduce maintenance. The flowering steppe is a landscape model inspired by the flat vegetation of garrigues where sheep roam on the stony plateaux of Mediterranean mountains, exposed to wind, cold and drought. In areas of the garden subject to occasional foot traffic one can recreate a flowering steppe by planting a mixture of allelopathic groundcover plants on a bed of gravel, together with, for example, light grasses and various bulbous plants, to create a planting that requires minimal maintenance. In our garden we have planted crocuses, narcissi and the grass *Stipa barbata* in the spaces between the patches of carpeting plants dominated by *Psephellus bellus* (syn. *Centaurea bella*) and *Tanacetum densum*, two remarkable groundcovers native to the steppes of Anatolia. In spring our steppe is covered in butterflies when the psephellus produces its abundant

flowers, while in summer the silver patches of the tansies give rhythm to the space, remaining ornamental in spite of drought. In autumn the magnificent flowers of the saffron crocus (*Crocus sativus*) take over as the drought breaks, with their multitude of pale mauve corollas that surround the bright red stigmas emerging directly from the soil. In winter it is the turn of pure white narcissi (*Narcissus papyraceus*), with scented flowers that faithfully light up the garden every year at Christmas time. This covering of the soil by alternating stretches of gravel and carpets of allelopathic plants limits the evolution of the vegetation, making a landscape which is in equilibrium, resistant to drought and adapted to foot traffic, and which moreover requires very little maintenance.

The gardens that Jennifer Gay and Piers Goldson create on islands in the Cyclades are often visited by their owners only during holiday periods. The use of a large range of garrigue plants with allelopathic properties, such as cistuses, phlomises, rosemaries, artemisias, Cretan thyme (*Thymbra capitata*, syn. *Coridothymus capitatus*), pink savory (*Satureja thymbra*) and Syrian oregano (*Origanum syriacum*), makes it possible to reduce maintenance in the garden over long periods of absence when the plants have to look after themselves. An allelopathic planting can also be the best solution when areas need to be planted that are hard to access or where maintenance might be dangerous.

Saffron crocuses (*Crocus sativus*) flowering in early November between clumps of *Psephellus bellus* and *Tanacetum densum*.

Top: To reduce maintenance, this border beneath trees is home to a mixture of allelopathic plants including *Cistus creticus, C. × argenteus, C. × florentinus, Ballota acetabulosa, Lavandula × ginginsii, Origanum syriacum* and *Phlomis × cytherea*. The garden of Vilka and George Agouridis, designed by Jennifer Gay and Piers Goldson.

Bottom: The allelopathic properties of the golden Coolatai grass (*Hyparrhenia hirta*) allow use on a large scale with little maintenance. Garden designed by Thomas Doxiadis on the island of Antiparos.

An allée on the top of Richer de Belleval's Mount in the Jardin des Plantes at Montpellier. The north-facing slope has been invaded by luxuriant vegetation consisting of sclerophyllous plants growing between the trees. On the south-facing slope the gardeners have opened up the landscape, using heliophile plants such as collections of cistuses and phlomises.

The outside edge of the green roof on top of the new Athens Opera House forms a projecting corbel tens of metres above the ground. This linear space, forbidden to gardeners for safety reasons, is planted with a double band of allelopathic plants that make weeding unnecessary: a border of grasses (*Hyparrhenia hirta*) accentuates the long fringe of prostrate rosemaries that spill over the edge of the roof and hang down over the void. *Hyparrhenia hirta* has also been used by the landscape designer Thomas Doxiadis in one of his most spectacular designs on the island of Antiparos, which recreates in the garden the golden colour of the agricultural land that stretches along the coast: over several thousand square metres the supple grasses, constantly swayed by the wind, make a sea that surrounds the house on all sides. The allelopathic properties of *Hyparrhenia* mean that a very limited amount of maintenance is required.

• *Reorienting the landscape: maintenance when an environment becomes closed*

The 'Mount' is one of the oldest parts of the Jardin des Plantes at Montpellier. Created by Richer de Belleval, who established the garden at the end of the 16th century, the Mount is about 135m (443ft) long, 24m (79ft) wide and some 3m (10ft) high, and is oriented east-west in order for its two slopes, north- and south-facing, to provide the conditions needed by plants from different habitats: plants from the coast and the driest garrigues on the southern side, and plants from shadier and cooler habitats on the north. Originally conceived as a teaching aid, to introduce students from the Faculty of Medicine to botany through observation of a collection of wild Languedoc plants, Richer's Mount was without doubt one of the very first gardens to be inspired by the diversity of garrigue habitats. As a result of various periods during which the Jardin des Plantes was partially abandoned, the Mount was for a long time left to its own devices: over the centuries, trees and sclerophyllous shrubs completely overran the old terraces.

To rehabilitate the Mount, the gardeners of the Jardin have recently cleared several south-facing terraces. Garrigue plants have once again been planted in the sunniest zones – collections of cistuses and phlomises, various plants that evoke the phrygana of the Eastern Mediterranean, a collection of coastal plants, and an original display of garrigue plants named after the city of Montpellier, such as *Aphyllanthes monspeliensis*, *Astragalus monspessulanus*, *Cistus monspeliensis*, *Coris monspeliensis* and *Acer monspessulanum*. The north-facing slope of the Mount has been left as it was, without any interventions; here a luxuriant and almost impenetrable vegetation has developed, concealing the ruins of the stone walls that supported the old terraces designed by Richer de Belleval. The dense foliage of viburnums, phillyreas, lentisks, buckthorns, Balearic boxwood and kermes oaks blocks the view, the plants jostling each other and growing tall to seek the light between the trunks of bay trees, holm oaks and hackberries which have self-seeded everywhere. Ivy, acanthus and butcher's broom (*Ruscus aculeatus*) infiltrate any space available beneath the trees, while osyris (*Osyris alba*) and redoul (*Coriaria myrtifolia*) occupy the few spots with more light. The contrast between the south-facing slope, reworked by the gardeners to create a new garden consisting of heliophile plants, and the north-facing slope where the vegetation has completely closed the environment, reminds us of the diverging trajectories that the evolution of the different zones of a garrigue garden may follow.

Drawing inspiration from the natural dynamics of plants in the various habitats found in Mediterranean land-

Top: The garrigue garden becomes richer as it matures, with the evolution of its different areas leading to a biodiverse environment that is home to an ever-richer fauna. The garden created by Jennifer Gay and Piers Goldson in the village of Rou, Corfu, restored by the architect Dominic Skinner.

Bottom left: Lying in wait on the flowers of *Achillea cretica*, a crab spider catches a dronefly which had come to feed on nectar. Insects, which soon arrive in a young garrigue garden, are the first link in the food chains that will be established in the garden.

Bottom right: Suspended from an esparto grass (*Lygeum spartum*) stem, a praying mantis slowly emerges from its sloughed skin.

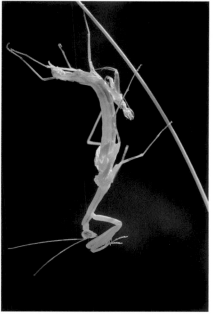

scapes, the gardener may create complementary zones in which the selection of plants and the type of maintenance practised will lead these parts of the garden in different directions. Some zones can remain open in the long term, adapted to the amount of foot traffic they receive. Others can be planned to develop into dense plantings consisting of a diversity of heliophile sub-shrubs, where an occasional clipping and selective removal of self-sown seedlings enables the area to remain in an equilibrium phase. And finally, yet other zones can be planned to evolve into taller plantings, where myrtles, viburnums, buckthorns, phillyreas, lentisks, box, junipers, bupleurums and privet (*Ligustrum vulgare*) will in time dominate the space with their evergreen foliage.

The garrigue garden becomes richer as it ages; the evolution in time and space of its different zones leads to a landscape that is home to an extensive range of plants and an ever-richer fauna. Insects rapidly move into a young garrigue garden: from the very first years the diversity of plants rich in pollen and nectar and the successive flowering periods, including in autumn and winter, attract a cohort of pollinating insects which are the first links in rich and complex food chains that will be established in the garden. A diversity of birds is reached more slowly. At the beginning, the open spaces in a young planting attract few species. Although a robin may follow the gardener as he or she digs during preparation of the soil, searching the newly turned earth carefully, the real ballet of garden birds starts only when there are taller sclerophyllous shrubs in parts of the garden with dense foliage offering nesting sites and berries providing an additional source of food – most small garden birds having a double diet, eating insects or fruits depending on the time of year.

In a planting where shrubs are growing taller, the rate of its evolution accelerates. A new kind of maintenance may then become necessary. When conditions are favourable for their propagation, sclerophyllous plants seek to become dominant. These plants self-seed profusely as soon as

perching spots are available to birds at the minimum height that guarantees their safety, from where they disperse seeds efficiently in their frequent droppings. Zones of low-growing garrigue, where the seeding of plants by berries scattered by birds is rare, are easy to preserve in an equilibrium phase with little maintenance. However, zones planted with trees and beds of taller shrubs that offer perches, whether these are evergreen sclerophyllous shrubs or large deciduous shrubs, such as chaste tree (*Vitex agnus-castus*), Sicilian sumac (*Rhus coriaria*), turpentine tree (*Pistacia terebinthus*), blackthorn (*Prunus spinosa*), spindle (*Euonymus europaeus*), smoke tree (*Cotinus coggygria*) or bladder-senna (*Colutea arborescens*), have a strong probability of being colonized by young sclerophyllous plants which will grow to provide perching spots in their turn, thus speeding up the cycle of closing an environment. When the garden is being planned, attention needs to be paid to the positioning of trees and shrubs that will provide perches, as the abundant scattering of seeds by birds will in time necessitate maintenance to control the profuse vegetation. At ground level it is often the common wild madder (*Rubia peregrina*) and wild asparagus (*Asparagus acutifolius*) that first indicate the propagation of berry-bearing plants by birds, followed – if these plants grow within a radius of a few hundred metres – by viburnums, honeysuckle (*Lonicera etrusca*, *L. implexa*), buckthorn and shortly afterwards the whole rich array of sclerophyllous garrigue plants, among which, thorny and invasive, brambles (*Rubus ulmifolius*) and sarsaparilla (*Smilax aspera*) move in, which can sometimes be extremely hard to eradicate.

If they have been planted as small specimens, garrigue trees develop on the same time scale as large sclerophyllous shrubs. Indeed, the distinction between large sclerophyllous shrubs and small trees is not always clear in the garrigue: depending on local conditions, shrubs sometimes grow taller than trees in the interval between two major disturbances, such as fire or felling, which start the vegetation off on a new cycle. In their first 10–15 years,

the evergreen foliage of arbutuses, mock privet (*Phillyrea latifolia*) and holm oaks blends into the same layer as lentisks, buckthorns, myrtles and narrow-leaved mock privet (*Phillyrea angustifolia*). When making a garden most gardeners of course don't have the patience to wait while garrigue trees slowly grow to the point where they provide shady areas protecting other plants from too much sun. The solution is usually to plant trees already grown on to a decent size in nurseries, which offer their beneficial shade much faster and add structure to the space. The range of trees that can be bought fully grown is limited, but planting less common complementary trees as small specimens or growing them in situ from seed – a subtle pleasure enjoyed by gardeners in the know – makes it possible to increase the diversity of the future tree layer

in a garrigue garden. Storax (*Styrax officinalis*), Cretan maple (*Acer sempervirens*), mock privet (*Phillyrea latifolia*), azarole (*Crataegus azarolus*), Saporta's pistachio (*Pistacia × saportae*), the Cyprus arbutus (*Arbutus andrachne*) and its hybrids that have become famous in the history of gardens, *A. × andrachnoides* and the Villa Thuret arbutus (*A. × thuretiana*), are magnificent trees which add richness to a garrigue garden. As soon as they reach a height of about 1.5–2m (5–6½ft), these young trees start to play the same role in the dynamic evolution of the garden as sclerophyllous shrubs, contributing to the closure of the environment and serving as perches for birds.

In areas of the garden that are progressively being dominated by sclerophyllous shrubs and young trees there are

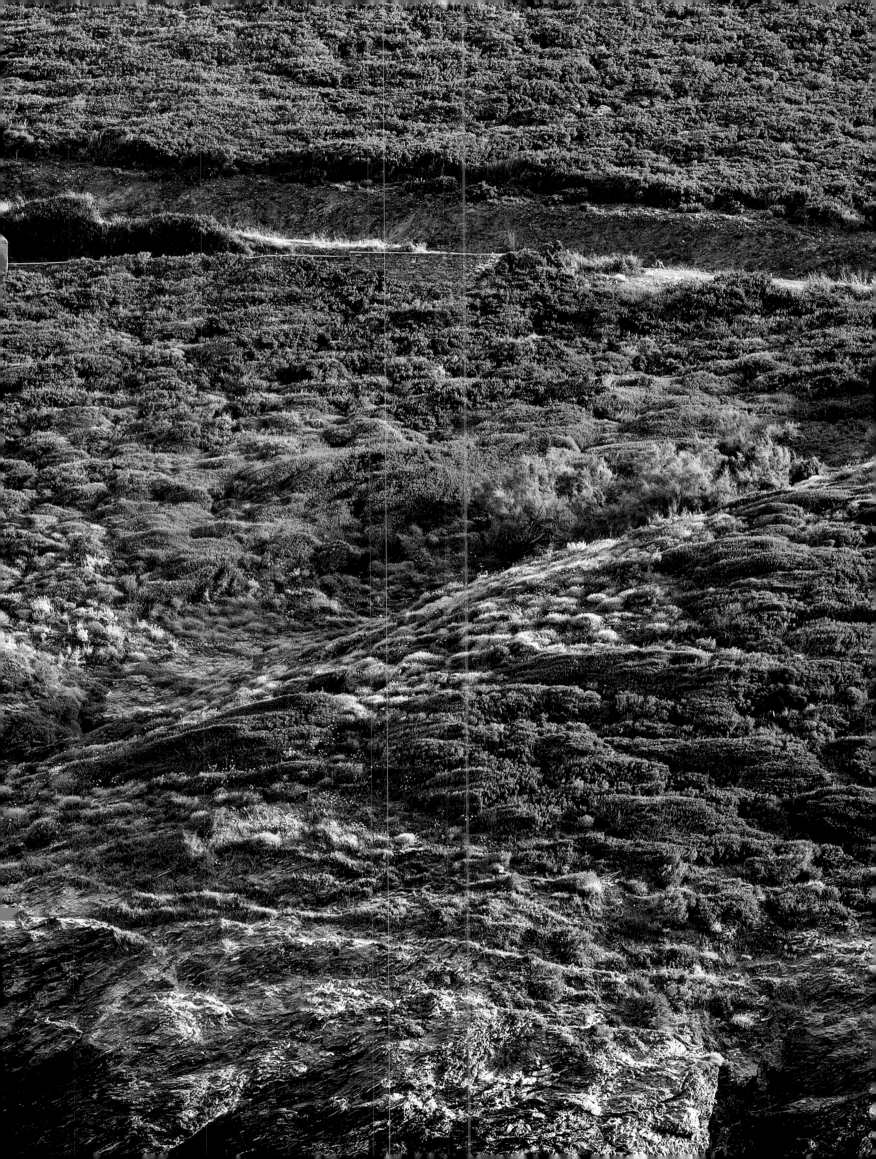

The magnificent trunk of an *Arbutus × thuretiana* twenty years after it was planted as a young plant only tens of centimetres high. Although it requires patience, planting unusual trees enables the gardener to achieve remarkable diversity in what will in future become the tree layer of the garden.

several maintenance options: either one can leave part of the garden to become closed and to develop into a wild planting, which serves as a refuge for fauna and which, if this planting is a large mixed hedge acting as a windbreak or screening a view, can be controlled by an occasional pruning; or one can intervene more drastically to turn this area into an open environment once more by removing some of the plants that have become undesirable. This may involve considerable labour, as the rootstock of sclerophyllous plants is deeply anchored in the ground; or one can open up areas between the layers of vegetation, in which taller plants are cleared away and a lower layer of sclerophyllous plants is left, kept low by regularly clipping.

Because of their long lifespan and tireless ability to sprout from the rootstock, the gardener can sculpt the vegetation of sclerophyllous plants into all sorts of heights and shapes. The old lentisks that form large round patches on the Cap des Trois Fourches in Morocco, the phillyreas that grow as rock-hugging cushions on Cap Creus in Catalonia, the myrtles and carobs whose foliage intermingles in fine folds on the Lycian coast of Turkey, the kermes oaks sculpted by goats on the slopes of Mount Zas on Naxos, or the Mediterranean buckthorns (*Rhamnus lycioides* subsp. *oleoides*) that form perfect spheres on the Gramvoussa peninsula in Crete – all these are models that suggest how some of the gardener's sclerophyllous plants can be pruned to keep their shape tight and low.

The two-layered structure serves both aesthetic and functional purposes. In the Park of Saleccia in Corsica, the free silhouettes of old lentisks, phillyreas, oaks, olives and

cypresses emerging from a lower layer consisting of an undulating mass of regularly clipped sclerophyllous plants create an exceptionally beautiful rhythm – a striking example of garden art inspired by the wild vegetation of the maquis. To the visual attraction of this kind of scene is added the major fact of the reduced maintenance it requires: clipping some of the sclerophyllous plants – easy to do – is an effective way of blocking the dynamic development of the vegetation, which in some cases might become hard to control and no longer correspond to the gardener's wishes. When they are not clipped regularly, sclerophyllous plants become so dense that they leave no room for competition. Here, sclerophyllous plants with allelopathic properties, such as lentisks and myrtles, reveal their full usefulness to the gardener: the density of their clipped vegetation which completely prevents light from reaching the soil, coupled with the powerful germination-inhibiting action of the chemical compounds diffused by the plants, offer the gardener a simple way of modulating the evolution of the landscape.

The compact masses of regularly clipped lentisks leave no room for competition because they prevent light from reaching the soil and they diffuse allelopathic compounds. The Park of Saleccia, Corsica.

A GARRIGUE GARDEN OUTSIDE THE MEDITERRANEAN

The landscapes of Provence, Languedoc and Roussillon are the stuff of dreams. The alternation between golden wheatfields and perfectly aligned rows of lavender on the Valensole plateau; the thousands of flowering rosemaries that light up the garrigues on the Minerve hills in the depths of winter; the terraced vines of Banyuls which take on yellow, orange and russet hues in the magnificent autumn light; the warm resinous scent of pines in the calanques of Marseilles, or the still air vibrant with the sound of cicadas in the garrigues of Montpellier in early summer – all these images conjure up a gentle Mediterranean climate. Yet this gentleness masks a far more complex reality: the climate of garrigue landscapes can be subject to some brutal variations. Most garrigue plants are adapted not only to the heat and drought of Mediterranean summers but also to climate conditions that some people might find surprising in Mediterranean regions: torrential rain and occasional periods of sharp frost. This ability to tolerate a wide range of precipitation and temperatures

Top: *Euphorbia officinarum* subsp. *echinus* south of Agadir, growing on the artificial dam of the Youssef Ibn Tachfin reservoir, whose level varies from year to year. Rainfall on the southern side of the Mediterranean is characterized by great variability from year to year. In late November 2014 rain reaching as much as 250 mm (10in) locally fell in the space of 48 hours between Agadir and Ouarzazate, the equivalent in some areas to a year's worth of rain.

Bottom: The torrential rains that occasionally hit the arid landscapes of the Atlas cause spectacular erosion.

makes it possible to consider using many garrigue plants in gardens situated outside Mediterranean regions.

• *The effects of rain*

In the Mediterranean climate, the annual rainfall is characterized by its irregular distribution throughout the year. The dry period of summer is often followed by heavy downpours in autumn. Extreme weather conditions in the Mediterranean can refer to rain and floods just as much as heatwaves, drought, wind and fires. In the winter of 2015, when we were on a botanic trip to study the pre-desert garrigues of southern Morocco, Clara and I were struck by the extent of visible damage to the infrastructure of roads and bridges. Near the wadis, roads had disappeared and bridges were broken or altogether collapsed: a bumpy track led across dry riverbeds where the road had become impassable. A few months earlier, after four years of terrible drought that had killed countless almond trees in the high valleys of the Anti-Atlas, torrential rain suddenly hit southern Morocco. At the end of

November 2014, 250mm (10in) of rain fell in the space of 48 hours between Agadir and Ouarzazate, the equivalent in some areas of the normal annual rainfall. However, even this did not break the record of 17 August 1995, when an estimated 200mm (8in) of rain fell in two hours in the High Atlas, causing killer floods in the Ourika valley near Marrakech.

Top: In spite of its placid appearance, the River Vidourle, which flows through vineyards and garrigue landscapes from the Cévennes to the sea, can sometimes give rise to catastrophic floods. In periods of drought the discharge of the Vidourle is low, in the region of 10 litres (17½ pints) per second, whereas during violent storms in late summer or autumn it can be more than 1,800 cubic metres per second, in other words comparable to the average discharge of the Rhone at its mouth.

Bottom: The Roman bridge over the Vidourle at Sommières. In the first century CE it carried the road connecting Nîmes to Toulouse. Since antiquity the Vidourle has been known for its sudden spates: each pier of the bridge is protected by a spur at its base and each is also pierced by a window to allow the passage of water when the river is high. This bridge can be completely submerged during exceptional spates.

Heavy rain is also frequent in the garrigues behind Nîmes and Montpellier. Some villages situated by rivers are built to accommodate floods: in Sommières, the spates of the River Vidourle are so frequent that they have been given a name – vidourlades – and the lower part of the town expects to be flooded almost every year. In autumn and winter, when the south and south-east winds marin and grec blow, blustering for days on end, the moisture-laden air from the Mediterranean meets the first heights of the Cévennes. The clouds discharge their water vapour as rain and these recurrent rains, known locally as épisodes cévenols, can be exceptionally heavy. The garrigue landscapes which a few months earlier seemed crushed by heat and drought are now subject to torrential rain, with every decade setting a new rainfall record. In the area between Saint-Gély-du-Fesc and Bel-Air, about 12km (7½ miles) north-west of Montpellier, for example, 280mm (11in) of rain fell in 48 hours between 29 and 30 October 1955, 200mm (8in) in 24 hours on 16 October 1920, and 135mm (5¼in) in one hour on 30 November 1910. The floods of October 2014 in the Gardon basin behind Nîmes caused major damage, the épisode cévenol this time being a real deluge, with 303mm (12in) of rain falling in 12 hours in Uzès and 315mm (12½in) in only a few hours at the Garrigues military camp, a few kilometres from Nîmes. Already, 12 years earlier, on 8 and 9 Septem-

Top: A plaque on a house in Sommières showing the water level during the flood of 9 September 2002. In the lower town streets were under more than 4m (13ft) of water.

Bottom: *Euphorbia spinosa* in the Dinaric Alps, near the Velebit massif in Croatia, where the average annual rainfall is more than 1,500mm (60in). Garrigue plants can tolerate high rainfall provided the soil has perfect drainage, as here in a karst soil from which water drains rapidly.

ber 2002, the Department of the Gard had experienced the most terrible floods in its recent history, with an amount of rain that is hard to imagine in the drought-adapted garrigue landscapes of the South of France: more than 400mm (15¾in) fell in 24 hours in the catchment area of the Gardon, and in the Anduze region as much as 687mm (27in) in 24 hours – in other words more than the total annual rainfall of London, concentrated in a single day.

Garrigue plants are capable of tolerating both a lack of water and a temporary excess of it, for the history of garrigue soils has created an environment where water can drain away rapidly. A succession of torrential rains over many thousands of years has led to substantial erosion of land denuded by fires, by clearing and by grazing, resulting in the rocky, poor and well-drained soils in which garrigue plants thrive today. It is not surprising that most of the garrigue species are ill-suited to land that remains wet for a long time: an increase in the volume of soil taken up by water corresponds to a decrease in the volume taken up by oxygen. After rain, garrigue plants must rapidly find once more, between the large particles in the soil, the oxygen required by the complicated symbioses that exist between their roots and fungi and bacteria. Whether they are growing in the wild or in gardens, heavy or frequent rains cause no problem to garrigue plants as long as their roots are in soil that has perfect drainage.

• *The effects of cold*

For millennia, garrigue landscapes have experienced episodes of extreme weather that have left their mark. In regions where the winters can be harsh, the plants have adapted to tolerate occasional periods of severe cold, with the vegetation undergoing natural selection by recurrent exceptionally hard winters. During the 20th century, for example, the winters of the years 1929, 1956, 1963 and 1985 left their mark on the landscapes of the South of France, redrawing the boundaries of the zone where olives can be cultivated. The best-known episode is that of February 1956, for it was characterized by a double phenomenon particularly difficult for plants to withstand: sap rose very early, due to mild weather at the beginning of the year, but this was followed by the sudden arrival of intense cold which lasted for the whole of February. On 2 February 1956 temperatures dropped dramatically, in some places by more than 15°C (27°F) in 24 hours. Throughout France, polar temperatures set in, with front-page pictures in newspapers: in Paris the frozen Seine resembled an ice field, in Toulouse people put out tables and chairs on the Canal du Midi which was covered in a thick layer of ice, near Marseilles fishing boats were fixed to the spot when the brackish water of the Étang de Berre froze over completely, and at Aigues-Mortes the pink flamingos died where they stood, their feet trapped in the ice on the saltings. A glacial front from the northeast, caused by the dual effect of an anticyclone over continental Europe and low pressure in the Western Mediterranean, gave rise to a violent mistral, with gusts of more than 150km (93 miles) per hour sweeping down the Rhone valley. The cold spell lasted until 27 February, with extremely low temperatures across the whole of the South of France: –9°C (15.8°F) at Toulon, –11°C (12.2°F) at Perpignan, –16°C (3.2°F) at Montpellier and –20°C (–4°F) at Aix-en-Provence. In the Saint-Martin-de-Londres basin, which is often the coldest place in the Hérault Department, the temperature went down to –26°C (–14.8°F), similar to that recorded in Moscow during the same period. The cold wave of February 1956 was accompanied locally by significant snowfall, as for instance in the Alpes Maritimes Department and on the Var coast. There were 35cm (13¾in) of snow at Vence and as much as 70cm (27½in) at Saint Tropez, where strong winds created snowdrifts more than 1m (3¼ft) high. In the botanical garden of the Villa Thuret at Antibes, the soil remained covered by a thick layer of ice for more than a fortnight.

The occasional very low temperatures experienced by garrigue landscapes are not limited to the South of France. In cold spells of recent years, extremely low temperatures were recorded in southern Spain, for example –14°C (6.8°F) at Granada in January 1987 and –24°C (–11.2°F) at Albacete in winter 1971. In February 1956, the temperature was –10°C (14°F) by the sea in Mallorca and –13°C (8.6°F) at the monastery of Lluc in the Serra de Tramuntana, a high point in the botanical diversity of the garrigues of the Balearic Islands. In Italy, the cold spell in 1956 was accompanied by a violent blizzard, with heavy snowfall in Rome and Naples and the temperature dropping to –32°C (25.6°F) on the Fucino plateau in the Abruzzi Mountains above Rome. At the same time snow fell all over Sicily and even reached the island of Lampedusa, lying between Malta and the Tunisian coast. The cold waves and snowstorms did not spare North Africa either. In February 1956 it was –5°C (23°F) at Constantine in north-east

Top: The exceptionally cold spell in February 1956 redrew the zone around the Mediterranean in which olive trees can grow.

Bottom: *Bupleurum spinosum* on the slopes of M'Goun in the Atlas Mountains. In Mediterranean regions where winters can be very harsh, plants that are adapted to drought can also withstand recurrent spells of extreme cold.

Top: A landscape in southern Morocco, near the village of Tafraoute in the Anti-Atlas, at the end of January 2005. In the foreground the thorny branches of a wild jujube tree (*Ziziphus lotus*) bend under the weight of the snow.

Bottom: After heavy snowfall in Mallorca in February 2012 the broken branches of a holm oak lie on the ground and garrigue sub-shrubs are completely buried under the snow.

Algeria and −12°C (10.4°F) at Setif in Little Kabylie, while it snowed for 14 consecutive days at Ain Draham in north-west Tunisia.

These cold spells are not just distant memories. In January 2005 a great wave of cold swept over southern Spain and North Africa, with the temperature dropping to −8°C (17.6°F) at Córdoba, while Algiers was under snow for several days. At the end of January 2005, Clara and I were on a botanic trip in southern Morocco and were cut off by the snow for several days in the village of Tafraoute, for the cold spell had reached the hills of the Anti-Atlas, only a few kilometres distant from the large expanses of the Hamada desert. A few years later the month of February in 2012 was equally icy all around the Mediterranean, with temperatures of −4°C (24.8°) at Athens, −7°C (19.4°F) on the Ligurian coast near Genoa, and as low as −18°C (0.4°F) at the foot of Mont Sainte-Victoire near Aix-en-Provence, while there were major snowfalls in the Balkan peninsula as well as in Corsica, Sardinia, the Balearics and North Africa.

• *Microclimates*

During cold spells around the Mediterranean there may be significant differences in temperature within a region, depending on microclimates linked to the topography. The Saint-Martin-de-Londres basin, situated in the heart of the garrigues about 30km (19 miles) north of Montpellier, is famous for its extremely low winter temperatures. This

basin is a depression closed in on one side by the Hortus Mountain and the Pic Saint-Loup, and on the other side by the ridge which separates it from the Hérault valley, water from the basin escaping down a narrow canyon, the Ravin des Arcs, to join the Hérault River. This depression traps the cold air that flows down the slopes of the Hortus and the Pic Saint-Loup in winter, bringing temperatures that are sometimes 10–15°C (18–27°F) degrees lower than those in the surrounding areas. The temperature can become exceptionally low when thermal inversion linked to nocturnal radiation is favoured by a clear sky and a carpet of snow on the ground, as was the case in the cold spell of 1963 when a temperature of –29°C (–20.2°F) was recorded at Saint-Martin-de Londres on 4 February 1963. In these extreme climatic conditions, the position of garrigue plants in the landscape depends on their hardiness. The downy oak (*Quercus pubescens*) and the turpentine tree (*Pistacia terebinthus*) are able to survive in the icy bottom of the basin, while the holm oak (*Quercus ilex*) and the lentisk (*Pistacia lentiscus*) have to live a little higher up on the slopes in order to escape the most severe cold. The same phenomenon can be seen south-east of Aix-en-Provence, where the massifs of Mont Sainte-Victoire, Mount Aurélien, Regagnas and the Étoile trap cold air in the basin of the Arc valley; there, sub-zero temperatures are often similar to those recorded at Saint-Martin-de-Londres, with for example –23°C (–9.4°F) recorded at Pourrières in February 1956.

A great many Mediterranean coasts have high mountain chains running just inland from them. The range of garrigue plants changes according to distance from the sea and variations in temperature due to altitude. Although the temperature never reaches freezing point at Agadir in southern Morocco, the nearby mountains of the High Atlas can experience very harsh winters: the lowest temperature recorded in Morocco was –33.4°C (–28.12°F) in December 1934 at Tacchedirt, a village in the High Atlas south of Marrakech. In the 1930s, it was indeed on his observations of the groups of plants growing at various altitudes between the coast and the high mountains of southern Morocco that the botanist Louis Emberger based his definition of the different types of bioclimate in the Mediterranean. According to Emberger, it is summer dryness and not mild winters that characterizes the Mediterranean climate, with significant differences in winter temperatures at different altitudes. Mountains are just as emblematic of garrigue landscapes as coastal areas: the massifs of the Atlas and the Rif in Morocco, the Tellian Atlas and the Aures massif in Algeria, the Betic Cordilleras in Southern Spain, the eastern end of the Pyrenees, the Corbières massif, the Causses plateaus and the foothills of the Alps in the South of France, the Apennines which run the length of Italy, the massifs of Olympus, Pindos, Parnassus, Taygetos and Parnon in Greece, the Taurus mountain chain in Turkey, the Alawite mountains and Mount Lebanon running along the eastern shores of the Mediterranean, and the Jebel Akhdar in Cyrenaica – all these mountains are an integral part of the Mediterranean floristic region and many garrigue plants are able to grow at a high altitude where temperatures can be very low in winter.

As one climbs the Têt valley in the western Pyrenees, the highest outposts of common garrigue plants such as common thyme (*Thymus vulgaris*) and the Nice euphorbia (*Euphorbia nicaeensis*) can be seen in heathland covered

Inflorescence of a turpentine tree (*Pistacia terebinthus*) in early spring. The turpentine tree and the downy oak (*Quercus pubescens*) both grow in the coldest areas of the garrigues in the South of France, such as in the Saint-Martin-de-Londres basin.

Mount Dirfi on the Greek island of Euboea. Mountains are as much a feature of Mediterranean landscapes as coastal areas. Here, on the apparently bare slopes, a great variety of garrigue plants adapted to high altitudes grow, such as *Stachys euboica*, *Nepeta argolica* and *Origanum scabrum*.

in butterflies at an altitude of more than 1200m (3937ft). Higher still, near the fortified town of Mont-Louis, *Cistus laurifolius* colonizes any free space on the slopes facing the sun at almost 1500m (4921ft). Common thyme also grows on the Larzac plateau, together with a range of plants that can withstand the rigorous climate of the Causses, where temperatures of about −10 to −15°C (14 to 5°F) are not uncommon in winter: *Lavandula angustifolia*, *Helichrysum stoechas*, *Coronilla minima*, *Centranthus lecoqii*, *Geranium sanguineum*, *Origanum vulgare*, *Helianthemum apenninum*, *Euphorbia seguieriana*, *Teucrium aureum*, *Aphyllanthes monspeliensis* and *Phlomis herba-venti*.

However, not all garrigue landscapes around the Mediterranean are subject to occasional very low temperatures.

Because of their geographical situation, some areas have always enjoyed a milder climate, even during the severest cold spells. In the winter of 1956, for example, the Atlantic coast of southern Portugal, protected by the maritime influence, experienced relatively mild temperatures; indeed, the south-westernmost point of Portugal, the Cape Saint Vincent area, was one of the few parts of southern Europe where the thermometer did not fall below zero. Along the Atlantic coast of south Morocco, the Macaronesian landscapes, also benefiting from a maritime influence, are completely frost-free. In the magnificent garrigues between Safi and Tiznit on the Moroccan coast there can be found some tender species that also grow in the Canary Isles, such as the tree houseleek (*Aeonium arboreum*) and King Juba's euphorbia (*Euphorbia regis-jubae*). Peninsulas on islands, which benefit from a thermal

Top: Among oaks and box, the meagre grasslands of the Larzac plateau are famous for the richness of their flora. Many garrigue species tolerate the harsh winter climate of the Causses, including *Aphyllanthes monspeliensis*, the golden germander (*Teucrium aureum*) and common thyme (*Thymus vulgaris*).

Bottom: After late frosts, common in the Causses until April, the grasslands on the Larzac plateau are covered in flowers in May before they are dried out by the summer drought. Here, *Helianthemum apenninum*, *Euphorbia seguieriana* and *Thymus dolomiticus* near the hamlet of Mas Raynal.

buffer since they are almost entirely surrounded by expanses of sea, such as the peninsulas of Akamas and Karpas in Cyprus or the Akrotiri, Rodopou and Gramvoussa peninsulas in Crete, are also regions of the Mediterranean Basin where frost is rare.

Minimum winter temperatures can sometimes differ significantly within a very short distance. The accidents of topography can enable garrigue plants to benefit from sheltered conditions right beside much colder zones. South-facing cliffs and enclosed gorges opening on to the sea, for example, create microclimates exploited by various tender species. One of the most northerly sites of the tree euphorbia (*Euphorbia dendroides*) is at the foot of the impressive Dévenson cliffs in the calanques of Marseilles. Myrtle (*Myrtus communis*) is rarely seen in

King Juba's euphorbia (*Euphorbia regis-jubae*) on Cape Rhir, south of Essaouira. The landscapes of the Macaronesian domain along the Atlantic coast of southern Morocco are protected from frost by the maritime influence.

the garrigues of Languedoc, where the climate is usually too cold for this warmth-loving shrub. However, there are fine populations of it in the Clape massif near Narbonne, where it thrives in the shelter of the cliffs above the Rec watercourse. In south-west Crete, the many gorges that cut through the southern slopes of the White Mountains also create highly localized microclimates. Walked by thousands of tourists every summer, the Samaria Gorge begins at the level of the Omalos plateau and comes out almost 1250m (4100ft) below on the shores of the Libyan Sea. At the end of the gorge, near to the little port of Agia Roumeli, one may see growing among the pebbles the small golden cushions of the rare plant *Teucrium cuneifolium*, a tender species that is found only in the most sheltered areas of the south-western coast of Crete. Above the Samaria Gorge, by contrast, may be found species that are extremely resistant to cold such as *Sideritis syriaca*, which can withstand temperatures in the order of –20°C (–4°F) without problem, according to the hardiness tests carried out at the Prague Botanical Garden (see box on p.229).

• Drought resistance helps frost resistance

Some garrigue plants are able to withstand extremely low temperatures. But how do these plants, native to the Mediterranean, manage to do it? In numerous garrigue species there is a partial convergence between their strategies to resist drought and their strategies to resist cold. This convergence involves both the morphology of the plant and complex biological mechanisms which enable it to avoid frost damage within its cells. In winter, during extended periods of frost, the plant's water equilibrium is disturbed: having become partially frozen in the soil and in the sap-conducting vessels, water is no longer very available to the plant. Yet its leaves and stems continue to lose water through evapotranspiration. This loss of water can be particularly high for evergreen plants when the cold is combined with strong winds, as is often the case in the harshest winters around the Mediterranean. The plants thus suffer from intense hydric stress, but the very same strategies by which they withstand drought help garrigue plants to get through cold periods without hydric deficit.

The extensive root system that enables garrigue plants to use the little water available in the soil during the long dry periods of summer also reduces their vulnerability in winter to the drought brought about by the freezing of the soil surface. The structure of the leaves may also play a part in limiting hydric stress caused by the cold. The glossy cuticles that cover the leaves of sclerophyllous plants, the protection of the stomata beneath a thick coat of hairs, and a reduction in leaf size are all common strategies to enable plants to resist hydric deficit, whether caused by drought in summer or by frost in winter.

Some alpine plants adopt strategies comparable to those of garrigue plants. The edelweiss (*Leontopodium nivale* subsp. *alpinum*) and the various wormwoods found high up in the Alps (*Artemisia glacialis*, *A. umbelliformis*, *A. umbelliformis* subsp. *eriantha* and *Artemisia genipi*) have leaves that are covered in silver hairs, like numerous garrigue plants. In alpine plants this layer of hairs reflects back excessive solar radiation in summer and also limits water loss in periods of frost by trapping a layer of still air which acts as a thermal insulator to protect the underlying tissues. A number of garrigue plants that grow in mountains show a considerable similarity to silver-leaved alpine plants, using the same strategies to withstand both drought and cold. The mountains of Greece, for example, are home to a great diversity of Mediterranean plants with silver leaves that have become adapted to live at high altitudes. *Convolvulus boissieri* and the golden-flowered heron's bill (*Erodium chrysanthum*) on Mount Parnassus, *Stachys candida* and *Scabiosa crenata* on Taygetos, *Achillea umbellata* and *Cerastium candidissimum* on Mount Parnon, the velvet horehound (*Marrubium cylleneum*) and the tiny germander with large blue flowers (*Teucrium aroanium*) on Mount Aroania in the Peloponnese – all have leaves covered in silver hairs to withstand drought in summer and cold in winter. The silver-leaved plant that is perhaps best known in gardens, lamb's ears (*Stachys byzantina*), was for a long time called *Stachys olympica*, evoking the mountain habitat of its place of origin, the

Bithynian Mount Olympus, or Uludağ, one of the mountains of Anatolia.

A ball- or cushion-shaped habit, a strategy adopted by numerous garrigue plants to resist drought and heat in the Mediterranean littoral, also enables them to keep their vegetation hugging the ground in the mountains where temperatures are slightly less cold. In high Mediterranean mountains, plants shelter each other from the icy wind by growing in a tight succession of cushions or as carpets, just as plants protect each other from salt on the coasts of Corsica and Sardinia. In the White Mountains of south-west Crete, the hikers' trail that goes up to the Pachnes massif enters the Amoutsera valley just after passing the relict forest of magnificent multi-centenarian cypresses. Beyond the last trees, which have taken on amazing twisted shapes to survive at 1650m (5413ft), the vegetation becomes ground-hugging in order to better withstand wind, cold and snow. *Berberis cretica* is transformed into spiny carpets, junipers (*Juniperus oxycedrus*) embrace the rocks to form a vast groundcover, their trunks twisted by the wind. Higher up in the Amoutsera valley, aromatic plants which remain buried in the snow for a long time, sometimes until May, often have a hemispherical shape, for example *Helichrysum italicum* subsp. *microphyllum* and pygmy savory (*Satureja spinosa*). Their cushion shape allows these plants to benefit from the thermal insulation under an even layer of snow, while the

Top: *Convolvulus boissieri* grows on many mountain massifs around the Mediterranean, for example on Mount Parnassus above Delphi and in the Sierra de Cazorla in southern Spain. Some Mediterranean plants that grow at higher altitudes adopt the same strategies to resist drought and cold as alpine plants, with silver leaves and a cushion-shaped habit.

Bottom: The leaves of lamb's ears (*Stachys byzantina*) are covered in long silver hairs. These hairs reflect back excessive solar radiation and reduce water loss during frosts by trapping a layer of air which acts as a thermal insulator to protect the underlying tissues.

anti-fungal properties of their essential oils protect them from fungal pathogens which can easily develop in the confined space beneath the snow.

Spiny xerophytes (from the Greek *xeros*, dry, and *phuton*, plant) are also ball-shaped plants, covered in spines, which often grow at high altitudes. Their slow-growing vegetation forms perfect topiaries, anchored by a powerful taproot. The long roots, tiny leaves and extremely dense vegetation and the way the tips of their stems and leaf petioles have become spines allow spiny xerophytes to withstand extreme conditions whatever the season: heat, drought, wind, cold or snow. Thus spiny xerophytes are found on countless Mediterranean mountains: *Genista lobelii* on Mont Sainte-Victoire in Provence, the blue *Erinacea*

anthyllis in the Sierra Nevada in Andalucia, *Astragalus angustifolius* on Mount Gingilos in Crete, *Dianthus webbianus* and *Acantholimon trojanum* on Mount Ida in north-west Turkey, *Bupleurum spinosum*, *Ptilotrichum spinosum* and *Arenaria pungens* in the high mountains of the Atlas. In winter on Mediterranean mountains, the whole carpet of snow sometimes lies in hummocks, reproducing the shapes of the plants below, with the rhythm of light on this undulating carpet of snow making an amazing winter variation of the Mediterranean landscape.

• *Natural antifreeze of the garrigue*

The morphological adaptations of Mediterranean plants, which may correspond to common plant strategies to avoid the effects of cold or drought, are often not enough to protect plants during periods of the most intense cold. Whatever their habitat, plants cannot survive if ice crystals form within their cells as this definitively ruptures their complex cell structure. Some garrigue plants have developed the same mechanisms as alpine or boreal species to avoid the formation of ice crystals in their cells when the outside temperature remains below zero for a long time. These mechanisms go through an acclimatization stage, which is triggered in autumn by the shorter days and cooler temperatures, and then speeds up with the advent of the first sub-zero temperatures. During these acclimatization stages the plants modify their metabolism to prepare to withstand the cold. For example, they increase the concentration in the cellular liquid of

Top: The Amoutsera valley in the Pachnes massif in Crete. Instead of developing into upright trees, junipers (*Juniperus oxycedrus*) form carpets flattened against the rocks by wind and snow.

solutes with an antifreeze effect, including soluble sugars and specific proteins which lower the freezing point of water. During a period of extreme cold, the ice crystals which start to form in the plant tissues are concentrated only in the empty spaces between cells, where the solute concentration is weaker. The ice crystals beginning to form outside the cells cause a difference in pressure on either side of the cell wall, leading to a movement of the water content of the cell towards the outside. Under pressure this water crosses the cell wall and increases the volume of the ice crystals in the intercellular spaces. If

the cold period lasts for a long time, the cells become progressively dehydrated, which means that they contain an ever greater concentration of their antifreeze solutes, lowering the freezing point even further.

During the concentration of the antifreeze solutes in the cellular liquid, some proteins play a complementary role in preventing the formation of ice crystals in the cell. In certain conditions water can remain liquid down to −39°C (−38.2°F): this is what physicists call the 'surfusion state'. By preventing the formation of the first ice crystals

Opposite bottom: Exposed to cold, snow and wind, centuries-old cypresses in the Amoutsera valley huddle into themselves, taking on the form of magnificent natural bonsais.

Top: Winter landscape on Puig d'Ofre in Mallorca. In Mediterranean mountains a carpet of snow sometimes forms humps and hummocks, moulding the shapes of the plants beneath.

Bottom: *Astragalus angustifolius* at an altitude of almost 2000m (6562ft) in scree on Mount Gingilos in Crete. Spiny sclerophyllous xerophytes are Mediterranean mountain plants that grow in spiny balls in order to better resist heat, drought, wind, cold and snow.

the antifreeze proteins allow the water contained in the cells to remain in a surfusion state, without freezing, even at very low temperatures. It is because of the concentration of antifreeze solutes in the cells that some garrigue plants are able to live in natural environments where winters can sometimes be harsh, whether in the high mountains of the Mediterranean or on plains with particular topographic features, such as the Saint-Martin-de-Londres depression or the Arc basin in the South of France, where temperatures can sometimes drop below −20°C (−4°F). The physiological mechanisms of adaptation to cold have a major disadvantage. Frost-hardiness, which requires a modification of the permeability of the cellular membranes and an accumulation of soluble sugars and antifreeze proteins, involves a significant energy cost for the plant. Only some garrigue plants acquired these specialized

biological mechanisms, because of the environment in which they live. The ability to tolerate frost is an adaptation to environmental conditions that has evolved over a very long time indeed, with the genetic coding for frost-hardiness being passed down from generation to generation. Thus garrigue plants show a great variation in their frost tolerance depending on their evolutionary history. A plant's frost tolerance may result as much from its long history of migration, on the geological time scale, during successive periods of glaciation and warming (see the following chapter), as from its more recent evolution in response to its local environment. Some garrigue plants thus retain the ability to withstand cold even when today they live in habitats which do not experience sub-zero temperatures. *Cistus ladanifer* var. *sulcatus*, which grows on the cliffs of Cape Saint Vincent in southern Portugal, is very hardy even though it is endemic to a restricted coastal area where in the current climate conditions temperatures never go below zero. *Cistus halimifolius* (syn. *Halimium halimifolium*), which colonizes the ground beneath the Barbary thuyas (*Tetraclinis articulata*) near the dunes of Essaouira in southern Morocco, grows in an environment where it never freezes. We collected seeds of this plant and have grown it for many years in an area of our garden dedicated to the study of a systematic collection of cistuses from all around the Mediterranean Basin. The pleasant surprise was that this cistus, which originally interested us for its lime tolerance, has proved to be a lot more tolerant of cold than one would have expected given its geographical place of origin: it withstands short periods of −10 to −12°C (14 to 10.4°F) without any damage at all.

However, another member of the Cistaceae family which grows in close proximity to *Cistus halimifolius* on the Essaouira dunes, the Canary Island sunrose (*Helianthemum canariense*), has a poor tolerance of cold, its main area of distribution being on the coast of the Canary Islands. The evolutionary history of plants now growing together in the flora on a particular site can in fact differ between species. Thus in the garrigues of the Essaouira dunes we can distinguish between species with affinities to the flora of the northern Mediterranean, such as *Ephedra fragilis*, lentisk (*Pistacia lentiscus*), *Phillyrea latifolia* and *Cistus halimifolius*, and plants with a more southerly origin, such as *Periploca laevigata*, *Ononis natrix* subsp. *angustissima* and the Canary Island sunrose.

• *Collections of hardy garrigue plants*

The aesthetic qualities of garrigue plants, their incomparable scents, their ability to do well in poor soils and their low maintenance requirements make them sought after by many gardeners, regardless of the climate zone they live in. Indeed, some of these plants have become so common that people forget they are native to Mediterranean garrigues. Lavender and common sage are sold on the banks of the Seine in Paris and in the garden centres of England and Germany, and the packets of narcissi, crocuses and anemones which tourists buy in the flower market beside the Singel canal in Amsterdam as typical souvenirs of Holland are for the most part cultivars of wild species from the garrigues of Turkey, Greece or Spain. Garrigue plants that regularly experience cold in their natural habitats can live without problem in gardens in regions where the winters are harsh, be it in the cold inland areas of the South of France or in more northern regions. It is, however, impossible to tell in advance whether garrigue plants that live in regions with mild winters will have a chance of withstanding cold. Their frost resistance depends on their evolutionary history: it can only be tested by exposing plants to cold conditions.

Many of the species so symbolic of Mediterranean garrigues, including cistuses, lavenders, rosemaries, oreganos, euphorbias, thymes, phlomises and sages, have been the subjects of trials in cold regions. Cistuses, for example, have long kindled enthusiasm in botanists and passionate gardeners: as a result, through the long history of cultivating cistuses in Europe, information has been amassed on the hardiness of most of the species and cultivars used in gardens. A relatively recent genus on the scale of plant evolution, cistuses express their potential to evolve rapidly by their tendency to hybridize in the wild. In the 18th

The degree of hardiness in garrigue plants depends on their evolutionary history. *Cistus halimifolius*, which grows in the dunes of Essaouira, tolerates cold well, whereas other species which grow in the same environment are much more sensitive to frost.

A heather (*Erica multiflora*) covered in snow in the Mourèze cirque in the Hérault Department of France. Garrigue plants that are regularly subjected to snow in their natural habitats will adapt well to gardens situated in regions with harsh winters.

century, botanists such as Linnaeus and Lamarck often believed that the hybrids they discovered in the garrigues of the South of France were new, hitherto undescribed, species. In order to understand the mechanisms of hybridization better, the botanist Édouard Bornet carried out research on cistuses at the end of the 19th century, establishing a huge collection of species and of hybrids created by manual pollination at the Villa Thuret in Antibes. To check the presumed parentage of natural hybrids, he reproduced by artificial fertilization cistus hybrids common in the South of France, such as *Cistus × pulverulentus* (which is a cross between *C. albidus* and *C. crispus*), *C. × hybridus* (a cross between *C. salviifolius* and *C. populifolius*) and *C. × stenophyllus* (a cross between

C. monspeliensis and *C. ladanifer*). In 15 years Édouard Bornet performed more than 3000 manual pollinations at the Villa Thuret: starting with 16 plants that he considered to be separate species, he ended up growing more than 200 distinct combinations. It was one of the very first major cistus collections in gardens.

From the gardener's point of view, one of the attractions of hybrid cistuses is their accumulation of the most interesting characteristics of their parents, such as lime tolerance and resistance to cold. Whether natural or created by artificial pollination, cistus hybrids are thus often easier to grow in the garden than their parents, for they are adapted to varying conditions. Already in the 1820s the

English nurseryman Robert Sweet was enthusiastic about cistus hybrids and minutely studied their behaviour in English gardens. In his work *Cistineae*, published between 1825 and 1830, he describes more than 30 cistus hybrids and encourages people to grow them in England for the beauty of their flowers and evergreen foliage, but also for their ease of cultivation and their excellent adaptation to local climate conditions. Some of the cistuses described by

Sweet are still regularly cultivated in gardens in Europe, for example *C. × oblongifolius*, a hybrid remarkable for the abundance of its lovely white flowers, with an excellent resistance to cold that comes from one of its parents, *C. laurifolius*. In the 1950s, Eric Sammons, an English amateur gardener, was passionate about cistuses and for long years conducted a meticulous programme of crossing cistuses whose beauty and hardiness he felt gave them major horticultural potential for English gardens. Some of the hybrids created by Sammons have become well known and are grown today in gardens not only in the South of France but elsewhere in Europe, for example *C. × argenteus* 'Peggy Sammons', with wonderful soft pink flowers with a metallic sheen, *C.* 'Ann Baker', with white flowers blotched with purple, which thrives today in the Botanical Garden of Prague where it withstands the cold perfectly, and *C.* 'Snow Fire' with magnificent blotches, one of the most widely cultivated cistuses in England.

In the 1980s large collections of cistuses were established in France, with the soil and climate conditions of each

Ornamental bark on an old laurel-leaved cistus (*Cistus laurifolius*). The laurel-leaved cistus, resistant to cold, is one of the parents of numerous cistus hybrids appreciated in northern countries for their excellent hardiness.

region providing additional data on the potential of cistuses to adapt to gardens. The collections formed by Gabriel Alziar in the Botanic Garden of Nice, René Échard at Prades in the Department of the Pyrénées-Orientales, Paul Pècherat in the Department of Charente and Albert Lucas in the Department of Finistère have all been carefully examined by the botanist Jean-Pierre Demoly, who is carrying out an in-depth study of the genus. In the spirit of Édouard Bornet, Jean-Pierre Demoly performs manual pollination to analyse the parentage of various plants whose origin is uncertain, such as *C. × lenis* 'Grayswood Pink', which is currently the hardiest of all the cistuses cultivated in gardens. At the same time, two reference collections have been established in the UK, one at the Chelsea Physic Garden in London, where more than a century earlier various cistuses supplied by Robert Sweet had already been cultivated, and the other

in Bob Page's garden in Leeds, in the north of England, where the climate can be very cold and damp. At the Chelsea Physic Garden almost 150 taxa, including species, sub-species, botanical varieties, hybrids and cultivars, are being studied. Bob Page's collection is even more extensive, progressively enriched by the results of his own manual pollination work. This garden illustrates in a spectacular manner the adaptive potential of some garrigue plants outside the Mediterranean zone: Bob Page's collection is without any doubt the largest collection of cistuses to have been assembled to date. Many people have benefited from this collection for their research work. It was Bob Page, for example, who provided the plant material used for the genetic sequencing of cistuses, carried out by Beatriz Guzmán at the Botanical Garden of Madrid, as well as for the trials of frost resistance made at the Research Center of the University of Oregon, near Portland in the United States. Bob Page also generously gave us several varieties which over the years have become solid assets in our nursery; one of his best crosses is *C. × pagei*, an abundantly flowering cistus which gained a tolerance of lime from one of its parents, *C. parviflorus*, and its excellent frost resistance from the other, *C. laurifolius*.

The fact that the richest collection of cistuses, a plant so symbolic of Mediterranean garrigues, is in the north of England might seem surprising – yet it is not the only outstanding collection of garrigue plants to be found in the UK. In the late 1990s, when Clara and I were looking into the species and natural hybrids of *Phlomis* in Turkey and Spain, we asked the advice of Jim Mann Taylor to help us organize our research trips. The author of a book describing all the *Phlomis* of the Mediterranean and Asia Minor, he had a magnificent collection of them in his garden near Gloucester in south-west England. Indeed, studying the behaviour of *Phlomis* in England is nothing new. Philip Miller, one of the head gardeners at the Chelsea Physic Garden in the 18th century, made observations on which *Phlomis* had survived the winter of 1740, reputed to have been one of the coldest in the past three centuries.

Top: *Cistus × lenis* 'Grayswood Pink' is currently the hardiest cistus grown in gardens: it can withstand prolonged periods of about –25°C (–13°F) without damage.

Bottom: In a garden near Leeds in northern England, Bob Page studies the behaviour of a large collection of cistuses. In the climate of the north of England cistuses have a markedly longer flowering period than in Mediterranean regions, with many varieties flowering non-stop throughout the summer. Photograph Bob Page.

Above: *Lavandula angustifolia* is hardy in the UK when it is cultivated in soil with good enough drainage.

Right: *Lavandula × chaytorae* 'Sawyers' is a hybrid whose parents are *Lavandula angustifolia*, native to the Alps of Haute Provence, and *Lavandula lanata*, native to the mountains of Southern Spain. It is appreciated in gardens in the UK for its compact habit, its woolly silver foliage and its deep purple flowers.

In more recent times, the famous British gardener Christopher Lloyd also studied some species from this genus. In his book *The Well-Tempered Garden*, he mentions the long life of *Phlomis fruticosa* in his garden at Great Dixter in south-east England, where after 45 years one specimen had formed a mass almost 3m (10ft) wide.

The UK has other surprises too for lovers of garrigue plants. Our study of artemisias took us to the garden of John and Jean Twibell near Cambridge, where during the first decade of this century they amassed a large collection of Mediterranean artemisias, including many species from Spain and Morocco. While in Cambridge we also visited the Botanic Garden, where lavenders have pride of place: Tim Upson, a world lavender expert, is the botanist

in charge of this garden. His book *The Genus Lavandula*, co-written with the botanist Susyn Andrews, gives precise information on the hardiness and cultivation conditions of more than 450 taxa of lavenders. Exceptional collections of euphorbias, thymes, oreganos, sages and rosemaries can also be found in the UK. All of these collections are valuable sources of information on the possibilities of growing garrigue plants outside the Mediterranean.

• *Garrigue gardens in damp climates: first and foremost, drainage*

The garrigue plants that are found in specialist collections and botanic gardens are already widely grown in gardens in the UK and the Netherlands. Most of these gardens, whose soil and climate conditions are very different from those of the plants' natural habitat, have one thing in common. The fundamental factor that allows garrigue plants to be grown outside the Mediterranean is the drainage of the soil. On his website (www.cistuspage.org.uk), Bob Page sums up how to grow cistuses in the north of England with three main pieces of advice:

1. Situation: a sunny situation, preferably on a south-facing slope or in a raised bed.
2. Soil: any well-drained soil. Cistuses will thrive in poor, stony or sandy dry soils. They are good for gravelled beds and part-paved areas.
3. Soil preparation: add no organic material or fertilizer of any kind. Incorporate a generous amount of gritty sand to improve drainage.

Other collectors of garrigue plants in northern Europe have come to the same conclusion: the cultivation advice they all give focuses on drainage. In the booklet on cistuses published by the Chelsea Physic Garden it is recommended that they be grown in beds slightly raised by the addition of gravel and sand. A few kilometres from Antwerp, in the cold, wet climate of north Belgium, an amateur gardener has applied this principle: in his collector's garden, open to the public, Achiel Vanheuckelom has for 20 years been growing in raised beds of sandy soil a selection of

cistuses, lavenders and rosemaries that tolerate temperatures that may sometimes go down to −15°C (5°F). In his
book on *Phlomis*, Jim Mann Taylor also insists on good
drainage, as do Tim Upson and Susyn Andrews, who
explain that many lavender varieties are perfectly able to
withstand cold provided that the soil is not soggy in winter.

In the UK, many nurseries have a demonstration garden
attached to their propagation area. Those of some nurseries
that specialize in botanical collections of garrigue plants
serve as models of drainage techniques. Jekka's Herb
Farm, Jekka McVicar's nursery near Bristol in south-west
England, is known for its collection of more than 500
aromatic and medicinal plants, of which many come from
Mediterranean garrigues: here one may find, for example,
collections of oreganos, thymes, lavenders, rosemaries and

savories. The demonstration garden of this nursery consists of a striking succession of raised beds, which not only
provide excellent drainage but also put the aromatic plants
within reach of visitors, who may touch them to discover
the diversity of their scents.

Roger and Linda Bastin's nursery, situated near Maastricht
in the Netherlands, also specializes in garrigue plants,
including many that are aromatic. Here the demonstration
gardens scattered through the nursery are crammed with
aromatic plants such as lavenders, sages, thymes, oreganos,
savories, artemisias, helichrysums, santolinas and rosemaries, along with a great variety of phlomises, ballotas,
cistuses, hellebores, euphorbias, giant fennels, germanders
and mulleins – an opulence of Mediterranean plants which
thrive thanks to their planting in raised rock gardens.

Cistuses and
rosemaries in their
natural habitat on the
stony screes of Saint-
Guilhem-le-Désert in
the South of France.
The acclimatization of
garrigue plants in
regions outside
mediterranean-
climate zones is
usually possible only
in conditions where
soil drainage is
excellent.

Jekka's Herb Farm, a nursery near Bristol in south-west England, is known for its collection of more than 500 species and varieties of aromatic and medicinal plants, many of which are native to Mediterranean garrigues. The demonstration garden at Jekka's Herb Farm consists of a succession of raised beds to ensure good drainage.

The late British gardener Beth Chatto was a true visionary in her approach to planting. Passionate about plants and inspired by her husband Andrew Chatto's research on the ecology of plants in their natural habitat, she was one of the first people to develop an approach to gardening that was based on research into plants and their adaptations to their local climate and soil conditions. Her gardens and nursery are situated near Colchester, in one of the driest areas of eastern England: although summer showers are not uncommon, the rainfall in summer is low and the annual rainfall does not exceed an average of 550mm (21½in), in other words less than the annual rainfall of Montpellier or Marseilles. In the 1960s Beth Chatto made her first experiments with dry-climate plants, creating a garden inspired by Mediterranean landscapes on a south-west-facing slope near her house. In her first book, *The Dry Garden*, published in 1978, she explains how she studied the landscapes of the Causses and the garrigues of Mediterranean mountains to make a garden that was adapted to her local conditions, where cistuses, euphorbias, rosemaries, lavenders, artemisias and sedums grew.

In 1992, Beth Chatto embarked on an innovative project which was to become one of her greatest successes. By this time her nursery had become very popular and she needed to create a new car park in order to cope with the growing crowds of visitors. She decided to transform the old car park into a garden, and instead of removing the layers of gravel with which it was surfaced, she chose to exploit these apparently inhospitable conditions to experiment with the idea of a gravel garden. In it she grew an

Top: In Roger and Linda Bastin's nursery near Maastricht, demonstration gardens are scattered through the nursery showing possible ways of using aromatic plants such as lavenders, santolinas, sages, helichrysums and rosemaries in the climate of the Netherlands. Photograph Roger Bastin.

Bottom: An explosion of Mediterranean colours in the Netherlands: plants flowering in June in Roger and Linda Bastin's nursery where a raised bed shows off a collection of lavenders and santolinas. Photograph Roger Bastin.

Top: *Allium sphaerocephalon*, *Sedum sediforme* and *Scabiosa columbaria* near the village of La Couvertoirade on the Larzac plateau. Beth Chatto was inspired by the landscapes of the Causses and Mediterranean garrigues to create a garden adapted to the conditions near Colchester, one of the driest regions of England.

Bottom: By studying the ecology of plants in their natural habitat, Beth Chatto developed an approach to gardening based on research into plants adapted to her soil and climate conditions. Here is a scene from her garden that includes cistus, lavender, euphorbia, asphodeline, valerian, fennel, nepeta and giant fennel.

extraordinary range of plants, including many garrigue species which in the wild live in a stony environment, so that they found here the soil and drainage conditions that suit them. This highly original handling of a former car park, in which Beth Chatto paid careful attention to the play of the shapes and textures of dry-climate plants, has become one of the best-known gardens in the UK. Since the publication in 2000 of her book *The Gravel Garden*, such gardens have been seen as one of the best landscaping techniques for growing garrigue plants in gardens outside the Mediterranean.

With more than a million visitors per year, Kew Gardens is one of the most visited gardens in the world. In one section of its huge rock garden there is a remarkable collection of plants from Mediterranean garrigues, of which many species are still rare in horticulture so that visitors are often able to admire them for the first time. The raised structure of this rock garden creates ideal drainage conditions that allow plants from Crete, Mallorca and Sicily to thrive only a few kilometres from central London. The mixture of earth, sand and gravel that fills the cracks between the large blocks of stone enables the

Making the most of stony and poor soil, Beth Chatto created a vast gravel garden on what was once a car park. Gardening on gravel is one of the best ways of growing garrigue plants outside mediterranean-zone climates.

Top: In Kew Gardens visitors may see a collection of garrigue plants including little-known species such as *Lomelosia albocincta* from the mountains of Crete.

Bottom: The raised structure of the rock garden at Kew creates drainage conditions that suit garrigue plants.

roots to grow down into an environment which always remains well-aerated, even in periods of frequent rain. Having visited this garden on and off for almost 30 years, Clara and I felt a little emotional when we saw at Kew some of the most beautiful plants that we have been able to observe in the wild during our travels round the Mediterranean: the blue-flowered *Petromarula pinnata* and the white-flowered peony (*Paeonia clusii*) from the mountains of Crete; the Balearic clematis (*Clematis cirrhosa*) and *Dorycnium fulgurans* from Mallorca; the woolly lavender (*Lavandula lanata*) and the lavender-

Top: *Euphorbia spinosa, Stachys byzantina, Hypericum olympicum* and *Thymus vulgaris* in Kew Gardens. In the rock garden at Kew the plants' roots are able to grow downwards into an environment that remains well-aerated, even in periods of constant rain, thanks to the mixture of earth, sand and gravel that fills the spaces between the large blocks of stone.

Bottom: Plants from stony garrigues live for a remarkably long time in regions outside the Mediterranean. Here we see an old specimen of the magnificent silver-leaved *Rosmarinus* × *mendizabalii* at Kew.

leaved sage (*Salvia lavandulifolia* subsp. *vellerea*) from the mountains of Andalucia; *Bupleurum spinosum* and *Phlomis bovei* subsp. *maroccana* with its bi-coloured flowers from the Atlas mountains; and *Origanum laevigatum* and the magnificent shrubby silene (*Silene fruticosa*) from the mountains of Cyprus.

We noted that some garrigue plants age particularly well in the climate of London. It is at Kew Gardens, for example, that the oldest specimen we know of the magnificent silver rosemary grows, the rare *Rosmarinus* × *mendizabalii*, which is native to the sea cliffs of Andalucia. Other garrigue plants too have attained a respectable age at Kew: *Lomelosia minoana* has for years formed a heavy cascade between the blocks of stone, an old specimen of the Anatolian *Onosma albo-roseum* has spread to become a large groundcover on the gravel, *Sideritis trojana* has formed an impressive mass of pure white downy foliage, and *Erinacea anthyllis*, which is reputed to be very slow-growing, has formed a thick cushion covered in violet-blue flowers every June.

Top: Founded in 1673, the Chelsea Physic Garden has a long history of growing Mediterranean plants. *Lavandula × allardii*, a hybrid of *Lavandula latifolia*, from the South of France, and *Lavandula dentata*, from southern Spain and Morocco, does well in the heart of London thanks to the urban microclimate.

Bottom: In Bristol Botanic Garden in south-west England a demonstration garden is dedicated to garrigue plants; here some key elements of the Mediterranean landscape can be seen – a range of shades of grey, green and silver as well as contrasts in the textures and heights of evergreen plants.

Kew is not the only botanic garden to contain collections of garrigue plants. Founded in 1673, the Chelsea Physic Garden, situated by the Thames in the heart of London, has a long history of growing Mediterranean plants. The rock garden there is home to some treasures from garrigues, such as the Atlas germander (*Teucrium musimonum*), which forms a silver cushion covered in bright pink flowers, the shrubby *Dianthus fruticosus*, which one can

see flowering in summer on the cliffs of the Rodopou peninsula in Crete, and the silver-margined phlomis (*Phlomis leucophracta*), which colonizes the rocky slopes by the road between Manavgat and Seydişehir in the western Taurus Mountains of Turkey.

The Botanic Garden of Bristol is a demonstration garden devoted to garrigue plants. Panels placed along the paths explain the history of Mediterranean vegetation, the strategies adopted by plants to withstand drought, and the potential use of these dry-climate plants in British gardens. This Mediterranean garden is situated on a long south-facing slope, off which water can run easily after rain. The top of the slope is planted with a dense bed of evergreen plants, myrtles, tree germanders, bupleurums,

Left: Acanthuses, hellebores, mulleins, nigellas, *Asphodeline* and *Ampelodesmos* compose this scene in one of the largest demonstration gardens for dry-climate plants at Hyde Hall, situated near Chelmsford north-east of London.

Bottom: The aromatic plants grown in raised beds at Hyde Hall include rosemary, lavender, hyssop, sage, oregano, thyme and fennel.

Sideritis syriaca, from the mountains of Crete, and *Thymus dolomiticus*, from the Causses in the South of France, grow in an experimental area in the Prague Botanical Garden where the hardiness of Mediterranean plants is studied.

coronillas, holm oaks and cypresses, while on the slope below, supported by rocks and low stone walls, is a succession of cushion- and ball-shaped plants, *Convolvulus cneorum*, *Anthyllis hermanniae*, *Euphorbia rigida*, *Teucrium marum* and *Satureja spicigera*, which create a supple rhythm of green, grey and silver shapes.

One of the largest demonstration gardens of dry-climate plants is Hyde Hall, to the north-east of London, near Chelmsford. Here, the entire terrain has been prepared to ensure perfect drainage: on a vast south-facing slope a mixture of sand and gravel has been incorporated into the soil to create ideal cultivation conditions for garrigue plants; after planting, the soil was covered with a thick layer of inorganic mulch. Planned as an educational space

where the visitor may discover a new range of plants that are undemanding and easy to grow, the dry garden of Hyde Hall comprises more than 400 species, including garrigue plants and species from the other mediterranean-climate regions of the world, which make a series of exuberant scenes mingling shrubs, perennials, bulbs, grasses and biennials which self-seed prolifically into the gravel.

● *A garrigue garden in a continental climate*

When Petr Hanzelka, in charge of the collections at the Prague Botanical Garden, contacted us for the first time, we were doubtful about whether it was worth introducing plants from around the Mediterranean into a cold region of Central Europe. But Petr Hanzelka was insistent,

In Prague Botanical Garden plants need to be able to tolerate a wide range of temperatures, the difference between minimum temperatures in winter and maximum temperatures in summer being up to 60°C (108°F). Among the many plants in this scene are *Salvia sclarea, Lavandula angustifolia, Phlomis lychnitis, Phlomis monocephala, Dorycnium hirsutum, Santolina insularis, Aurinia corymbosa, Salvia amplexicaulis, Salvia lavandulifolia* subsp. *oxyodon* and *Antirrhinum majus.*

taking the trouble to come from Prague to visit our nursery and discuss his project, namely to start a programme to test the hardiness of a wide range of garrigue plants which might be able to adapt to a continental climate. The climate of Prague is cold and dry, with an annual rainfall that does not exceed 480mm (19in). The temperatures of sometimes below −15 °C (5°F) in the Saint-Martin-de-Londres or Arc basins in the South of France made Petr Hanzelka smile: every winter Prague experiences a long period of bitter cold. But the sometimes burning-hot temperatures of the South of France in summer didn't daunt him either: Prague also has intense heat waves with temperatures that may be as high as, if not higher than, those in the South of France; the geographical position of the Czech Republic, far from the sea, means that there is

no maritime influence to temper the hot spells that come up from North Africa.

Petr Hanzelka and his team were motivated by the recurrence these days of heatwaves which are becoming more frequent. In the Prague Botanical Garden, the plants now need to cope more and more often with an extreme range of temperatures, the difference between minimum temperatures in winter and maximum temperatures in summer sometimes being in the region of 60°C (108°F). For example, at the end of January 2012, a cold spell settled over Central Europe. In Prague the nocturnal temperature went below the −20°C (−4°F) mark and in the south-west of the Czech Republic reached −38°C (−36.4°F). During this period the ground froze to a significant depth and

temperatures remained very low even in the daytime: from 2 to 7 February the daytime temperature did not rise above –7°C (44.6°F) in Prague, and temperatures remained steadily below zero for three weeks. But the year 2012 also saw an exceptional heatwave: during the great hot spell that extended over Central Europe in August, temperatures everywhere were over 35°C (95°F), with 40.4°C (104.72°F) recorded on 21 August at the Dobřichovice weather station a few kilometres from Prague. In August 2015 another heatwave hit the Czech Republic with temperatures nearing 40°C (104°F), higher than those recorded at the same time in many Mediterranean countries. The same scenario was repeated in 2017: a very cold winter with three consecutive weeks of freezing weather and temperatures going down to –25°C (–13°F) in February in the south-east of the Czech Republic, followed by a heatwave in August with temperatures once more in excess of 38°C (100.4°F) in the Prague area. Czech climatologists take the possible consequences of climate change very seriously: according to their projections, alternating cold and heat, once rare, may become the norm in the near future, within only a few decades.

In their search for species able to tolerate both heat and cold, Petr Hanzelka and his team at the Prague Botanical Garden are preparing for the future. Created at the end of the 2000s, the Mediterranean section of this botanical garden covers about 5000sq m (5980sq yd). On a south-facing slope, low stone walls were built and rockfill arranged to facilitate the run-off of water, the surface drainage being completed by an inorganic mulch consisting of pebbles, gravel or schist chips depending on the area. An important collection of plants native to the mountains of the Mediterranean and the steppes of Asia Minor was put in place, Petr Hanzelka having propagated from seed in the Garden's greenhouses many species that he obtained through the botanic gardens' exchange network. The Prague Botanical Garden's Mediterranean section is experimental: it is an elimination garden, of a kind rarely created on this scale, the aim of which is to observe which plants die in order to discover those that are capable of withstanding cold. The wide range of garrigue plants that survived the winters of 2012 and 2017 is already opening up new prospects for gardens that are subject to periods of cold in winter and drought or heat in summer.

After ten years, the species most resistant to cold from the genera emblematic of the garrigue, such as cistuses, lavenders, sages, santolinas, dorycniums, oreganos, thymes, savories, hyssops, germanders, valerians, artemisias, tansies, stachys, sideritis and mulleins, create a striking and very beautiful garden in a continental climate. The Prague Mediterranean garden requires little maintenance and its vegetation is evolving rapidly from year to year, as in a model garrigue garden. In spite of the severe weather conditions, it is one of the best examples of a garrigue garden that Clara and I have ever seen.

Every year the team at the Prague Botanical Garden notes the behaviour of new species under trial and establishes a list of the plants that have best withstood the cold. These lists are of major interest to us, for they have enabled us to expand the range of garrigue plants' resistance to cold in our database (see p.276). Exchanging information on garrigue plants cultivated in different climate zones is useful to all gardeners. The climate and gardens are changing, and the prospect of climate change, which depending on the region may take the form not of a milder climate but of an increase in extreme weather events, means that plants able to tolerate equally well cold, torrential rain, heat and drought are going to be more and more sought after. Among garrigue plants, those species with the best resistance to cold are ideal candidates for the gardens of the future.

Opposite: The Mediterranean section of the Prague Botanical Garden covers an area of about 5000sq m (5980sq yd). On a south-facing slope low stone walls have been constructed and rockfill used to help drainage. Here we see *Cistus laurifolius*, *Achillea coarctata*, *Tanacetum densum*, *Lavandula angustifolia* and *Hypericum olympicum*, as well as different species of thymes, helianthemums, sedums and micromerias.

1. *Thymus vulgaris* in a dry-stone wall in the Prague Botanical Garden.

2. In a cold climate *Lavandula angustifolia* and *Lavandula × intermedia* find ideal growing conditions at the top of a slope. Here they are surrounded by a collection of thymes and oreganos, *Hypericum olympicum*, *Linum suffruticosum* and *Linum campanulatum*.

3. A Mediterranean atmosphere at Prague Botanical Garden: in June hundreds of flowers open on *Cistus × cyprius* 'Troubadour', a hybrid of *Cistus laurifolius* and *Cistus ladanifer* var. *sulcatus*. The other plants in this scene include *Lavandula angustifolia*, *Santolina insularis*, *Achillea clypeolata* and *Rosmarinus officinalis* 'Arp', a rosemary cultivar that is particularly resistant to cold.

Garrigue plants in the Prague Botanical Garden

To illustrate the diversity of garrigue plants able to tolerate cold, the following list gives a small selection of the species which survived the winters of 2012 and 2017 without damage in the Prague Botanical Garden, with uninterrupted frost for several weeks and temperatures that sometimes fell below −20°C (−4°F).

Achillea clypeolata
Achillea coarctata
Anthyllis montana
Antirrhinum majus
Arbutus unedo
Artemisia santonicum
Asphodeline lutea
Asphodelus albus
Bupleurum fruticosum
Catananche caerulea
Centranthus ruber
Cistus 'Ann Baker'
Cistus laurifolius
Cistus populifolius
Cistus × cyprius
Cistus × hybridus var. *corbariensis*
Cistus × lenis 'Grayswood Pink'
Cistus × oblongifolius
Cupressus sempervirens
Dianthus pyrenaicus
Dorycnium hirsutum
Euphorbia characias
   subsp. *wulfenii*
Euphorbia myrsinites
Globularia meridionalis
Helianthemum apenninum

Hyssopus officinalis
Jasminum fruticans
Lavandula angustifolia (numerous
   cultivars)
Lavandula × chaytorae 'Richard
   Gray'
Lavandula × intermedia (numerous
   cultivars)
Linum narbonense
Marrubium supinum
Onosma cinerea
Origanum laevigatum
Origanum vulgare
Paliurus spina-christi
Phillyrea angustifolia
Phlomis bourgaei
Phlomis chrysophylla
Phlomis leucophracta
Phlomis monocephala
Phlomis samia
Pistacia terebinthus
Ptilotrichum spinosum
Rosmarinus officinalis 'Arp'
Rosmarinus officinalis 'Miss
   Jessop's Upright'
Salvia amplexicaulis

Salvia canescens
   var. *daghestanica*
Salvia lavandulifolia
   subsp. *lavandulifolia*
Salvia lavandulifolia
   subsp. *oxyodon*
Salvia lavandulifolia subsp. *vellerea*
Salvia multicaulis
Salvia officinalis
Santolina etrusca
Santolina neapolitana
Santolina rosmarinifolia
Satureja montana
Satureja spinosa
Satureja subspicata
Sedum album
Sedum ochroleucum
Sideritis syriaca
Spartium junceum
Stachys lavandulifolia
Tanacetum densum
Teucrium aroanium
Teucrium aureum
Teucrium chamaedrys
Teucrium flavum

1. *Anthyllis montana*, a species that is common in the Causses region of the South of France, can withstand the wide temperature differences that characterize a continental climate with their hot summers and icy winters.
2. Plants native to the steppes and mountains of Asia Minor are naturally adapted to significant differences between summer and winter temperatures. Here *Salvia canescens* var. *daghestanica* cascades over a dry-stone wall in the Prague Botanical Garden.
3. Cypresses, lavenders, santolinas, helianthemums, sages and valerian in one of the experimental areas of the Prague Botanical Garden. In view of climate change, garrigue plants that withstand cold as well as drought are ideal candidates for forward-looking gardens outside Mediterranean zones.

A GARRIGUE GARDEN OPEN TO THE WORLD

Page 230: The transition zone between an area of the garden that is more strictly maintained, which includes plants of varied origins, and a wilder area where garrigue plants evolve freely. The garden of Rosie and Rob Peddle in southern Portugal.

1. Since the 1980s the range of plants available to mediterranean gardeners has expanded. *Convolvulus sabatius*, native to Italy and North Africa, has now become a common garden plant.

2. What to plant in a garden? The choice of plants depends in part on the gardener's cultural perception of nature and gardens.

3. The flowering of species tulips in a garrigue garden is always a moment of happy surprise. Here the Persian tulip (*Tulipa clusiana*) in our garden.

4. *Salvia pomifera*, *Cistus × tardiflorens* and *Euphorbia ceratocarpa* have different geographical origins but go well together in gardens since they have the same cultivation requirements.

Which plants should be included in a garrigue garden? The question may seem simple, but the answers to it can vary according to numerous factors. The choice of the range of plants to be used depends not only on technical, aesthetic or functional criteria but also on gardeners' cultural perceptions of plants, nature and gardens. Since we began studying Mediterranean plants, Clara and I have noticed a change in gardeners' perceptions of plants and landscapes. When the first plant fairs were organized in France in the 1980s, gardeners hoped for a more diverse choice of plants that could be used in the garden. In southern Europe, gardeners had already blazed new trails by growing a great many species which had hitherto been little known, sharing their experiences in books that became the first important works of reference on gardening in a Mediterranean climate. In the 1970s at Sparoza, the garden near Athens which today is the headquarters of the Mediterranean Garden Society, an association with international members, Jaqueline Tyrwhitt was passionate about using the wild Greek flora in the garden. She described her successes and failures in a delightful book, *Making a Garden on a Greek Hillside*. In Mallorca, Heidi Gildemeister tried out plants in her garden, which consists of a remarkable collection of species from different parts of the world. In one of her books she described the unusual experience of making a garden on a hillside frequented by sheep, enabling her to make a systematic study of plants that withstand grazing.

During the same period, in the South of France, Pierre and Monique Cuche created an experimental garden dedicated to the study of a collection of plants coming from the mediterranean-climate regions of the world. They were interested in the flora of the Mediterranean Basin and made, for example, an in-depth study of the plants of the mountains of southern Spain. It was Pierre who first told us of the extraordinary flora growing along the road leading up to the Suspiro del Moro pass in Andalucia (see p.56), and his eyes shone with infectious enthusiasm as he gave us seeds of *Ononis speciosa*, from which are descended all the specimens that now grow in our garden.

More recently, an interest in wild flowers and an emphasis on showcasing the local flora have enriched the creative dynamism of conceptual landscape designers. James and Helen Basson in France, Jennifer Gay and Piers Goldson in

5. Opposite: *Ononis speciosa*, which grows on the rocky slopes beside the road to the Suspiro del Moro pass in Andalucia, was introduced into gardens in the South of France by dedicated nursery owners Pierre and Monique Cuche.

Bottom: Interest in the wild plants of the Mediterranean has enriched the creative power of conceptual landscape designers. This garden on the island of Antiparos, created by Thomas Doxiadis, includes a mixture of local wild plants and non-indigenous plants which blend into the surrounding landscape.

Top: *Agave americana* and *Ampelodesmos mauritanicus* are two pioneer species which profit from the disturbed ground by the sea to colonize free space on the Amalfi peninsula in Italy.

Bottom: Sweet pomegranates, left, here the cultivar *Punica granatum* 'Fina Tendral', and the stigmas of the saffron crocus (*Crocus sativus*), right, are among the gifts that the garrigue garden offers in November. The pomegranate and the saffron crocus are both native to Asia but integrate perfectly into a garden inspired by the garrigue.

Greece and Marilyn Medina Ribeiro in southern Portugal, for instance, are all young landscape designers who make the most of their thorough knowledge of Mediterranean wild plants to create gardens that are largely inspired by garrigue landcapes. Many private gardeners are equally passionate about the diversity of plants from the Mediterranean Basin. Their gardens are enriched by collections of cistuses, euphorbias, artemisias, rosemaries, phlomises and germanders, while sclerophyllous plants such as lentisks, phillyreas, buckthorns, holm oaks and arbutuses have become common in gardens, often used to give structure so that the garden remains attractive in summer during the long dry period.

Following this period of research and the expanding range of plants in Mediterranean gardens, we have noticed a recent development. The initial interest in the use of local flora as a way of enriching the garden has shifted into a more restrictive attitude. The creation of gardens consisting solely of indigenous plants has sometimes taken on an ethical dimension, whereby plants are considered desirable or undesirable not in terms of their uses, appearance or behaviour, but in terms of their place of origin. The 'local flora' labels, initially intended to encourage the use of plants grown from genetically local stocks as part of a specific effort aimed at the ecological restoration of their natural habitats, have been wrongly interpreted as suggesting that preference should be given to indigenous plants for use in green spaces. The confusion arising from

media hype about supposedly invasive plants has also led to anxiety about plants from other places, with non-indigenous plants becoming seen *a priori* as potentially dangerous. In this context, there's a potential hazard in highlighting gardens inspired by the garrigue. It could be seen as supporting the exclusive use of indigenous species, which would be in complete contradiction to the history of Mediterranean landscapes. Garrigues have always evolved over time and space. The distinction between indigenous and non-indigenous plants is based on a narrow spatio-temporal scale which bears little relation to the dynamics

The Nice euphorbia (*Euphorbia nicaeensis*) colonizes the ruffes of Salagou in the South of France. The history of species migration around the Mediterranean gives us a better insight into the future of Mediterranean plants and gardens in the context of climate change.

Top: Vestiges of cultivation on the island of Amorgos. Around the Mediterranean, the interweaving of cultivated and wild landscapes, permanently remodelled by humans, forms a unique cultural landscape.

Bottom: Vines, olives and in the distance phrygana browsed by goats on the island of Naxos.

Opposite top: A landscape in southern Sicily between Palazzolo Acreide and Caltagirone: wheat cultivated between almond, citrus and olive trees. The introduction of plants over millennia has significantly shaped the cultural landscapes of the Mediterranean.

Opposite bottom: Human influence on the landscape in the mountains of southern Morocco: only a few traces of garrigue still exist, the landscape having slowly become desertified by the daily cutting down or pulling up of every tiny green shoot that could serve as fodder.

of the vegetation of the Mediterranean Basin. If we are to understand better what is at stake when we choose plants for gardens inspired by the garrigue, it may be useful to reflect more broadly on the fascinating history of the vegetation of the Mediterranean: its recent history, inextricably linked to humankind, which has shaped an exceptional cultural landscape all round the Mediterranean; its more distant history, arising from an incredible dynamic of mixings and cross-migrations, connected to successive changes in the climate; and finally its deep history, expressed in the surprising links between some of the wild plants we see in our own environment and plants living in other parts of the world. This review of the history of landscapes will also enable us to have a better understanding of questions about the future of Mediterranean plants and gardens in the context of forthcoming climate changes.

• *Cultural landscapes of the Mediterranean*

Around the Mediterranean Sea, interlocking jigsaw pieces of cultivated and wild land, full of plants of diverse origins and permanently shaped by humans, form a unique cultural landscape. Introduced plants, that sometimes become naturalized, have enriched the wild flora. Every year at the beginning of March a great almond festival is held in Tafraoute, in southern Morocco. Cultivated on the arid plateaux and in the high valleys of the Anti-Atlas, almond trees suddenly transform the austere landscape, where pink granite emerges from the sun-baked ground, into a mild and fragrant realm, burgeoning with flowers at the end of winter. In spite of the recent droughts which have damaged some of the orchards, there is currently renewed interest in almond cultivation: the almond festival of March 2017, for example, included a scientific seminar on the social and landscape role of the almond in arid mountain areas experiencing problems associated with climate change. Along with the argan tree (*Argania spinosa*), grown instead of almonds in the lower valleys of the Anti-Atlas, the almond is one of the factors that have led to socioeconomic changes in the mountain villages.

These changes can be seen, for instance, in the recent creation of many women's cooperatives, which have had a direct impact on the future of the landscape: the income generated by the renewed support for industries that process local products has reduced the extreme pressures on the garrigue of the last few decades. In the high valleys of the Anti-Atlas, between Tafraoute and Igherm, only rare remnants of garrigues remain, the landscape having slowly been turned into desert by the daily cutting down or pulling up of every minuscule green shoot that could serve as fodder. Here the future of the wild landscape is directly linked to that of the cultivated landscape: the emphasis on the traditional cultivation of the almond is contributing to the preservation of the wild flora of the ancient garrigues of the Anti-Atlas.

Native to Iran and Afghanistan, the almond tree (*Prunus dulcis*) was introduced into the Eastern Mediterranean in about 3000BCE. It was spread by the Greeks around the Mediterranean and introduced into North Africa in about 600BCE, where it met with soil and climate conditions

that allowed it to naturalize quickly. The almond grows spontaneously today in many regions around the Mediterranean. It has, for example, become a wild tree, one of many pioneer plants that colonize waste ground and roadsides around Montpellier where its white or pink flowers are appreciated as they herald the imminent end to winter. In a mediterranean climate the almond germinates easily after the first rains of autumn. In our garden, where its nuts are dispersed by squirrels, we allow it to self-seed freely to form a structural layer among the garrigue plants. On the first fine days of February we enjoy the sight of bees already hard at work indiscriminately gathering pollen

Following pages: At the end of winter in the high valleys of the Anti-Atlas the flowering almond trees transform an austere landscape into a gentle scented place.

Top left: Almond trees in the Sierra d'Almijara in southern Spain. Native to Iran and Afghanistan, today the almond is one of plants that symbolizes Mediterranean landscape.

Top right: All around the Mediterranean the flowering of almond trees is awaited as a joyful sign heralding the imminent end to the winter.

from almonds, rosemaries, French lavenders, tree germanders, groundsels, coronillas and euphorbias, which all offer an abundant supply of nectar and pollen in the depths of winter regardless of their place of origin.

The olive (*Olea europaea* subsp. *europaea* var. *sativa*), plant symbol of the Mediterranean, is one of the trees that has the closest links between wild and cultivated landscapes. It appears to have been first domesticated in the Middle East in 7000BCE, then in the Iberian Peninsula in about 4500BCE, as well as in other regions around the Mediterranean. The biologists Catherine Breton and André Berville have analysed the genetic origin of numerous olive varieties to trace their parentage and identify their history of migration from the different ancestral populations of the wild olive (*Olea europaea* subsp. *europaea* var. *sylvestris*). In the thousands of years of varietal selection, hybridization and the transporting of varieties from one region to another we see the whole history of the Mediterranean expressed in the origins of olive varieties, which reflect periods of trade, wars, conquests and human

migrations. The Phoenicians, for example, introduced olive varieties from Cyprus and the Eastern Mediterranean into Greece, Cyrenaica, Carthage, Spain, Portugal and Italy in about 1000BCE. Thus the varieties 'Poulo' in the Department of Pyrénées-Orientales, 'Giarrafa' in Italy, 'Amygdalolia' in Greece, 'Galéga' in Portugal and 'Chétoui' in Tunisia will all have been introduced from the same place of origin, Cyprus. In any given region the numerous local varieties grown today may have different origins, some of them deriving directly from a specific area where they were domesticated, while others have mixed origins, resulting from the hybridization of different lines. For instance, the variety 'Cailletier', cultivated in the Nice region, has its origins in Lebanon, Turkey and Israel, as does the variety 'Arbequina', cultivated in Catalonia and reputed to be not much affected by the olive fly. The variety 'Olivière', cultivated in the Departments of Aude and Hérault and producing very mild oil, has its origins in both the Eastern Mediterranean and the Maghreb. 'Négrette', cultivated in the Hérault Department, has its origins in Tunisia and Turkey. 'Picholine',

grown in the Gard Department and known for its deliciously aromatic and fruity oil, is also the result of crosses between original populations in the Eastern and Western Mediterranean.

After the complex history of its diffusion around the Mediterranean, today the zone in which the olive can be cultivated is larger than the distribution of the wild olive (*Olea oleaster*) from which it derives, some cultivated varieties being markedly more resistant to cold than the wild olive. The olives we see in the Departments of Hérault, Gard and Vaucluse are thus linked to humans, since the climate is too cold for the wild olive to survive here. A transition from cultivated to wild is nevertheless frequent, even in zones where the wild olive does not exist in the spontaneous vegetation: spread by birds, the cultivated olive naturalizes easily in garrigues. On the stony slopes of the Bout du Monde cirque at Saint-Guilhem-le-Désert, for example, olives self-seed around old abandoned terraces. Having turned into a spontaneous plant after its introduction by humans, the olive is becoming a new wild

plant, growing together with holm oaks and large sclerophyllous shrubs such as phillyreas, lentisks and buckthorns and contributing to the structure of a cultural landscape modified by humans.

On the margins of cultivated fields, olives self-sow and become wild among the sclerophyllous garrigue plants that colonize abandoned plots.

Trees are not the only landscape modifiers that are linked to human activities. Herbaceous plants transform Mediterranean landscapes in a way that is also associated with humans. The love-in-a-mist (*Nigella damascena*) and poppies (*Papaver rhoeas*, *P. argemone*) that brighten the margins of fields and the edges of garrigue paths originally came from the Fertile Crescent region of the Near East. Like most other messicole plants (from the Latin *messio*, harvest), they spread round the Mediterranean Basin with the expansion of agriculture in about 7000BCE. Once considered weeds, messicole plants have now become

heritage species subject to conservation measures: the regional nature park of Luberon, for example, encourages farmers to alter their farming practices to protect cornfield weeds such as cornflower (*Cyanus segetum*), pheasant's eye (*Adonis annua*), corncockle (*Agrostemma githago*) and shepherd's needle (*Scandix pecten-veneris*).

Another introduced herbaceous plant, of a much more imposing size, is the giant cane (*Arundo donax*), omnipresent in Mediterranean landscapes and native to East Asia. Its presence in the Mediterranean goes back to about

5000BCE. It has naturalized along ditches, on roadsides and in waste ground on the margins of garrigues, generally near water, and ever since its introduction has been used for multiple purposes: as stakes in vegetable gardens, as windbreaks around cultivated plots, as racks for drying fruit, as matting to shade pergolas, and as a light but solid material with which to make baskets, for example in Spain, or furniture in Morocco and Cyprus.

Arundo donax has also been inextricable from music since antiquity. The syrinx, or pan pipe, was an important instrument in ancient Greece and Rome, playing a part in various foundation myths of Western civilizaton. According to different sources it was the god Hermes who invented it, or the god Pan; in the latter version, the metamorphosis of the naiad Syrinx into a reed or cane to escape from Pan as he pursued her was followed by Pan's transformation of this cane into a set of pipes as he sighed over the loss of the beautiful Syrinx. A pan pipe made from the giant cane is on display at the Archaeological Museum of Naples. Excavated at Pompeii, it seems to be in almost perfect condition after being buried for almost two thousand years under the ash from the eruption of Vesuvius. Today the giant cane is used to make the reeds of many woodwind instruments. The Var region in the South of France is the world centre for producing reeds for clarinets, saxophones, oboes and bassoons as the local varieties of giant cane are reputed to make the best reeds, giving a sound of great quality.

Although the introduction of plants has to a great extent shaped the cultural landscapes of the Mediterranean, several other factors are responsible for profound upheavals in the structure and composition of ecosystems. The goats that sculpt kermes oaks and oleasters in the Cycladic islands or the sheep that graze on the meagre vegetation between Denmate and Ouarzazate in the Moroccan High-Atlas are not native to these regions. The goat (*Capra aegagrus*) and the sheep (*Ovis orientalis*) were first domesticated in eastern Anatolia in about 8500BCE. Their arrival around the Mediterranean followed the same wave of expansion as agriculture, between the seventh and fourth millennia BCE. Some sheep have reverted to their wild state, forming populations of mouflons in various regions, as for example the Troodos mountains in Cyprus, or near the cliffs of Cape Figari in north-west Sardinia, or in the Cinto mountains and near the Aiguilles de Bavella in Corsica. Domesticated goats, finding the conditions in Mediterranean mountains similar to those of their native habitat, have also sometimes reverted to the wild. In Crete there are protected populations of the wild goat, called in Greek agrimi or kri-kri, in the Samaria Gorge as well as on three small islands off the north coast, Dia, Theodorou and Agioi Pandes, where the

Bottom left: Love-in-a-mist (*Nigella damascena*), which colonizes the margins of cultivated land and the sides of paths in the garrigue, is native to the Near East.

Bottom right: The seed capsules of love-in-a-mist are full of small black seeds that germinate in poor soil that has recently been worked, which enables them to spread easily at the edges of cultivated land.

Top: A plain in the Fertile Crescent between Damascus and Aleppo in Syria. Poppies, love-in-a-mist and many other annual meadow plants come from the bare steppes that surround the plains where agriculture first developed about 10,000 years ago.

Bottom: Annual meadow plants that have escaped from fields on the Noto plateau in Sicily. Introduced plants that sometimes become naturalized have enriched the cortege of wild plants around the Mediterranean.

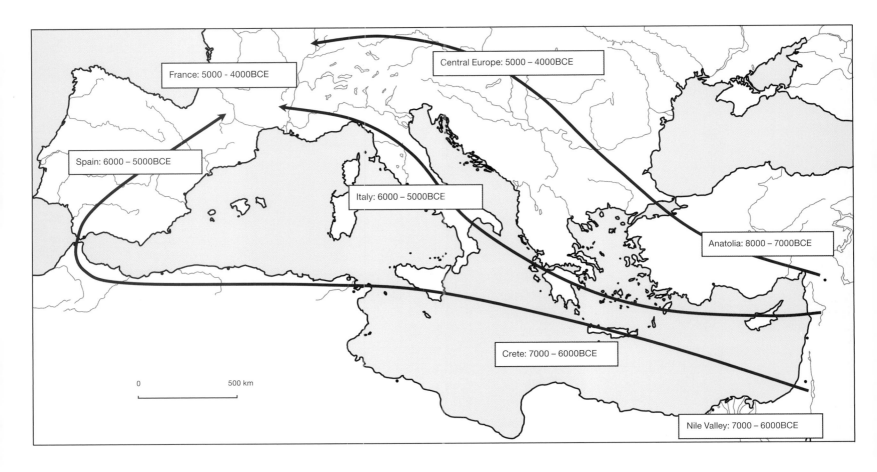

Top: Annual meadow plants migrated along the three main routes along which the cultivation of common wheat (*Triticum aestivum*) spread. Some are now considered heritage plants in the landscapes of the Western Mediterranean.

agrimi was introduced as part of a set of conservation measures. The impact of this goat today on the vegetation of these tiny islands, the stony and bare expanses of which call to mind some semi-desert islands off Croatia, reminds us of the profound influence that sheep and goats have had over the course of thousands of years on the cultural landscapes of the Mediterranean. Thus gardens inspired by garrigue landscapes, far from reproducing an ideal kind of nature, evoke the whole history of the way the landscapes around the Mediterranean have been transformed by humans.

• *Ice ages, migrations and the mixing of species*

A garden inspired by the garrigue is made up of plants which are the result of a long history of mixings and migrations around the Mediterranean – indeed, the history of migrations is part of the very essence of Mediterranean

Bottom left: Ever since its introduction into the lands around the Mediterranean, the giant cane has been put to multiple uses.

Bottom right: a pan pipe excavated at Pompeii is in almost perfect condition after being buried for nearly 2,000 years under the ash from the eruption of Vesuvius. Naples Archaeological Museum.

Top left: The goats that sculpt the vegetation in the Cyclades or Morocco are not native to these regions. Sheep and goats were introduced into the lands around the Mediterranean by man between the seventh and fourth millennia BCE.

Top right: A herd of goats on the slopes of Mount Dikti in Crete beneath Palestine oaks (*Quercus coccifera* subsp. *calliprinos*). The impact of grazing on the vegetation today reminds us of the profound influence that sheep and goats have had for millennia on the cultural landscapes of the Mediterranean.

Bottom: A shepherd and his flock in the Atlas Mountains, where over-grazing has completely changed the landscape.

landscapes. This history started long before cultural landscapes appeared that result from the impact of humans on nature. The Neolithic Revolution, which is known for the spreading of plant and animal species that had only recently been domesticated, occurred not long after the end of the last Ice Age, about 11,700 years ago. During the preceding geological era, four major ice ages, with a cold, dry climate, alternated with milder and wetter interglacial periods. These successive climate changes over a span of

almost 2.6 million years led to significant modifications in the distribution of the plants that now make up Mediterranean landscapes, as plants continually moved in response to changes in the climate in order to find conditions in which they could survive.

During the last Ice Age, which reached its peak about 22,000 years ago, the landscape of the South of France, for example, was very different from the one we know

Top: In gardens, the inspiration drawn from garrigue landscapes reflects the transformation of landscapes around the Mediterranean by humans. The garden of Camilla Chandon on Mallorca.

Bottom: Sheep maintain terraces planted with fig trees on the island of Naxos.

today. The glaciers of the Alps stretched as far south as Lyons, their ice having a thickness of more than 1km (3280ft). Further south lay a cold steppe, dominated by artemisias and ephedras, scattered with silver birches and *Pinus sylvestris*. Mediterranean species had taken refuge during this period in areas less affected by the glaciation, such as the southernmost points of the Balkan, Italian and Iberian peninsulas, as well as in North Africa. Occasionally some species managed to remain in place by hunkering down in sheltered microclimates, such as in coastal cliffs and valleys. During the interglacial periods, boreal species such as artemisias and birches moved towards more northerly latitudes, while Mediterranean species took advantage of the warmer conditions to reconquer the territories now freed from ice by the return of a milder climate.

Top: *Hyparrhenia hirta*, *Pistacia lentiscus* and *Phlomis × cytherea* in the park of the Stavros Niarchos Cultural Centre in Athens. A garden inspired by the garrigue consists of plants which are the result of a long history of mixings and migrations around the Mediterranean.

Bottom: Aleppo pines on the Dévenson cliffs in the calanques of Marseilles. During the ice ages, the sheltered microclimates of sea cliffs served as a refuge for some plants, while most mediterranean- or temperate-climate species migrated south to escape the ice.

During these significant migratory waves that accompanied the successive periods of cooling and warming, some plants proved to be markedly better than others at propagating themselves. The plants most common in the garrigue today are those species that have migrated repeatedly in the last few million years, sometimes over great distances, changing their distribution area more or less rapidly depending on changes in the climate. Several species now typical of the landscapes of the South of France are to be found all around the Mediterranean, in Lebanon, Morocco, Greece and Italy. Thanks to their effective seed dispersal strategies and their ability to adapt to different conditions, they managed to extend their distribution areas over the whole Mediterranean Basin during the interglacial periods, making use of wind, water, birds, insects and mammals. Species with a distribution all around the Mediterranean include mock privet

(*Phillyrea latifolia*), its fruits spread by birds, Montpellier maple (*Acer monspessulanum*), with keys (winged seeds) that are carried by wind and water, and Phoenician juniper (*Juniperus phoenicea* subsp. *turbinata*), the fleshy cones spread by birds and mammals. On the reduced scale of a garrigue garden, the pioneering behaviour that one sees in various species that self-seed easily, such as the lentisk (*Pistacia lentiscus*) and the mock privet (*Phillyrea latifolia*), corresponds to a remarkable potential to migrate over long distances, which has enabled these plants to change their distribution areas regularly in order to adapt to changes in the climate.

The distribution of other species around the Mediterranean is, by contrast, disjointed. In an earlier era they had a much wider distribution area which was subsequently fragmented by glaciation, and the plants have not man-

aged to regain their original distribution in any of the most recent migratory waves. Balearic boxwood (*Buxus balearica*) grows spontaneously in Mallorca as well as in southern Spain and Morocco. A few isolated populations are also found in the Taurus Mountains in Turkey. These distant distribution areas would have been connected in the past, at a time when the Balearic boxwood must have grown over the whole of the northern side of the Mediterranean Basin.

The geographical distribution of species can be very different within the same genus. The common arbutus (*Arbutus unedo*), for example, has a wide distribution all round the Mediterranean, as well as on the Atlantic littoral, while the Cyprus arbutus (*Arbutus andrachne*) is found only in the eastern half of the Mediterranean, and Pavari's arbutus (*Arbutus pavarii*) grows only in a very restricted

Cork oaks (*Quercus suber*) in the Alentejo in southern Portugal. During the glacial and interglacial periods, oaks constantly kept moving as the climate evolved in order to find conditions where they could survive.

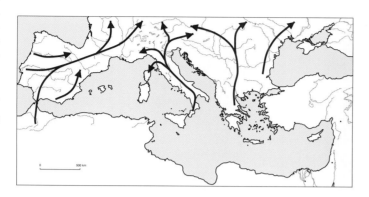

area in the mountains of Cyrenaica in Libya. In the successive glacial and interglacial periods, species in a particular floral group may have migrated in different manners, depending on their propagation method, their ecological requirements and their ability or not to cross any bioclimatic barriers, like mountains or the sea, that they might encounter on their way. The new assemblages of plants resulting from these varying migrations are entirely random. The species found together today in any particular Mediterranean region might have been geographically separated in the past, while conversely species which today are separated might have formed part of the same vegetation structure in earlier epochs.

Species with a scattered distribution illustrate the ability of garrigue plants to become integrated in different floral associations. The Cretan thyme (*Thymbra capitata*, syn. *Coridothymus capitatus*) grows spontaneously in the south of various Mediterranean peninsulas in Spain, Italy and Greece, but also in Morocco, Cyprus and Lebanon. At some point in its history it certainly had a much wider distribution area, of which the places where it is found today are only fragments. Looking at Cretan thyme growing on the shores of the Aegean Sea, one might believe it to be part of an ecosystem subject to overgrazing and salt spray, where its companion plants, all cushion- or ball-shaped to withstand salt and sheep's teeth, include heather (*Erica manipuliflora*), ballota (*Ballota acetabulosa*), broom (*Genista acanthoclada*), savory (*Satureja thymbra*) and spiny chicory (*Cichorium spinosum*). Yet when one sees Cretan thyme growing in the Sierra de Almijara, to the north of Almuñécar in Andalucia, it is obvious that it is part of an ecosystem composed of totally different genera

and species, including woolly lavender (*Lavandula lanata*), *Phlomis crinita*, Portuguese germander (*Teucrium lusitanicum*) and lavender-leaved sage (*Salvia lavandulifolia* subsp. *vellerea*). Cretan thyme growing in other environments where it is spontaneous, such as in Israel and Sicily, forms part of a completely different flora where its companion plants may be spiny burnet (*Sarcopoterium spinosum*), shrubby sage (*Salvia fruticosa*), *Prasium majus* or *Asphodelus microcarpus*.

Cretan thyme illustrates the aptitude of Mediterranean plants to grow with a wide range of other plants and in varied environments, as they have done naturally in the different phases of their migratory history. Of all the thymes that we love, Clara and I have always had a particularly soft spot for Cretan thyme, which we have studied all round the Mediterranean. We appreciate it not only for its ability to withstand the toughest conditions, but also because it is one of the plants that best represents our constant call to open the garrigue garden to its multiple zonal origins: in the garden it associates well with plants from different regions where it grows spontaneously as they share largely similar cultural requirements and live together very happily thanks to their modest demands for water and maintenance. The garden thus becomes a mirror reflecting what best characterizes the richness and diversity of Mediterranean plants, namely their ability to integrate with a wide range of plants. Plants move about, in gardens as in the wild. The way a community of plants is constantly renewing on the tiny scale of a Mediterranean garden recalls the uninterrupted history of mixing and recomposition on the scale of geological time.

Top left: Thanks to its fleshy cones that are easily spread by birds and mammals, the Phoenician juniper (*Juniperus phoenicea*) was able to exploit the flows and counter-flows of the flora during the glacial and inter-glacial periods to take over both sides of the Mediterranean.

Top right: On the reduced scale of a garrigue garden the pioneering behaviour of some species that self-seed easily, such as *Phillyrea latifolia*, corresponds to a migratory potential that enabled the plant to change its distribution area regularly in order to adapt to changing climates.

Bottom: A Phoenician juniper by the sea in the calanques of Marseilles.

Following pages: The Balearic boxwood grows spontaneously in Mallorca as well as in southern Spain and Morocco. A few isolated populations are also found in the Taurus Mountains in Turkey. These distribution areas, now far apart, may have been connected in the past when the Balearic boxwood must have grown throughout the whole of the northern Mediterranean.

• *Returning to the origins of garrigue flora*

The range of Mediterranean plants that can be used in gardens, and their close links with species growing in other Mediterranean-climate regions of the world, have their origins in epochs much earlier than the last ice ages: exchanges and migrations of Mediterranean-climate plants go back to the history of the continents themselves. Three hundred million years ago, all the land on earth formed a single continent, to which historians have given

Top: Cretan thyme (*Thymbra capitata*) forms part of various floral communities with different distribution areas around the Mediterranean, in Spain, Italy, Greece, Morocco, Cyprus and Lebanon. The plant communities in the garrigues, which are the chance result of migrations in the past, are at risk of altering in the future because of climate change.

Opposite, bottom:
The garden mirrors
what best
characterizes the
richness and diversity
of garrigue plants,
namely their ability
to become integrated
into different plant
communities
depending on soil and
climate conditions.
This scene includes
*Salvia sclarea, Salvia
officinalis, Cistus ×
pulverulentus* gr.
*Delilei, Santolina
insularis, Santolina
rosmarinifolia,
Thymus dolomiticus,
Thymus vulgaris,
Lavandula
angustifolia* and
Rosmarinus officinalis.

Top: A lavender-
leaved sage (*Salvia
lavandulifolia* subsp.
lavandulifolia) has
self-seeded in our
garden between a
Phlomis lychnitis and
a *Lomelosia minoana.*
The way a mix of
plants constantly
changes in a
Mediterranean garden
evokes, on a miniature
scale, the history of
the recomposition of
the Mediterranean
flora on the scale of
geological time.

the name *Pangaea* (from the Greek *Pan*, all, and *gê*, the earth). The ancestors of some of the plants from our garrigues already existed at the time of Pangaea. However, they lived then in an environment utterly different from the one they inhabit today: the climate was tropical and the flora was dominated by species that like heat and humidity. Arbutuses, oaks, hollies, yews, cypresses, buckthorns and pistachio species are all plants which originally grew in a tropical climate. Some of their biological characteristics, such as thick evergreen leaves and the ability to regrow from the rootstock, which initially were strategies to avoid being eaten by large herbivores, endowed these plants with a remarkable pre-adaptation to the Mediterranean climate that they were to encounter much later.

About 200 million years ago Pangaea began to split up. Two supercontinents appeared, one to the north called Laurasia, which included the future North America and Eurasia, the other, called Gondwana, to the south, which was later to fragment into different plates including those

that were to become Africa, South America and Australia. The Mediterranean Basin, situated at the point where the remains of Laurasia and Gondwana meet, today has plants that share a common origin with plants now growing in distant lands to the south. The tree heather (*Erica arborea*) that grows everywhere in Corsica and in areas of the South of France with siliceous soils, for example, is directly related to more than 600 species of heather growing in the floral kingdom of the Cape of Good Hope in South Africa. In our garden, one of the finest sclerophyllous shrubs with glossy foliage is the African olive (*Olea europaea* subsp. *cuspidata*). Similar to the Mediterranean wild olive, the African olive grows from Ethiopia to South Africa.

Sometimes the similarities in origin of plants now inhabiting different biogeographical zones are expressed in the relictual presence of an isolated species. The Barbary thuya (*Tetraclinis articulata*) is a magnificent conifer, the ancestors of which once grew throughout Europe but it is now found only in the Atlas Mountains and in a few small local populations in southern Spain, Malta, Tunisia

Top: 300 million years ago, all the emerging land formed a single continent to which historians have given the name Pangaea. The ancestors of some of our garrigue plants like the lentisk already existed at the time of Pangaea.

Bottom: Arbutus (*Arbutus unedo*) undergrowth above Porto Azzuro on the island of Elba. Arbutuses, oaks, hollies, yews, buckthorns and pistachio species are plants which originally grew in a tropical climate. Some of their biological characteristics such as thick evergreen leaves enabled these plants to adapt subsequently to a mediterranean climate.

and Algeria. The origins of the Barbary thuya predate the splitting of Laurasia and Gondawana: its closest relatives are species that grow today in Australia, such as the cypress pine with glaucous leaves (*Callitris columellaris*), or in South Africa, for example the Cederberg widdringtonia (*Widdringtonia wallichii*), often called the 'Cape cedar' although in fact it belongs to the cypress family.

One of the most important stages in the diversification of the mediterranean-climate flora happened about 65 million years ago, when Laurasia in turn began to split up, with the North Atlantic progressively opening to separate Eurasia from North America. The fragmentation of the distribution areas of plants which until now had grown side by side in Laurasia explains the amazing similarity between certain plants found today in the American Southwest and around the Mediterranean Basin. The Pacific madrone (*Arbutus menziesii*) is an arbutus that grows on the American West Coast. The first time that Clara and I saw a madrone in California we were struck by its close resemblance to the Cyprus arbutus; the magnificent trees growing in the mountains along the coast north of San Francisco could have been mistaken for those that

grow in the mountains along the coast in northern Cyprus: similar-looking leaves, the same splendid trunk with smooth red bark that peels once a year to reveal the underlying new green bark, the same ability to regrow from the rootstock after a disturbance, the same spring flowering with terminal panicles of creamy-white flowers followed by bunches of small red fruits in autumn. The Pacific arbutus and the Cyprus arbutus are vicariant species (from the Latin *vicarius*, taking the place of), in other words species which have remained very similar, arising from the breaking up of their common distribution zone.

The affinity between some of the flora of North America and the flora of the Mediterranean Basin can be seen in the number of genera that are common to the two regions: more than 50 genera of Mediterranean plants, for example, are found in the American Southwest. These genera include plants often used in gardens, such as sages, helianthemums, artemisias, atriplex, peonies, cypresses and oaks. Sclerophyllous oaks such as the coast live oak of California (*Quercus agrifolia*) and the holm oak (*Quercus ilex*) have many things in common: the shape of the acorns, the size of the acorn cups, the thick evergreen leaves with irregularly

Bottom left: A landscape of tree heather (*Erica arborea*) near Saint-Chinian in the South of France. The Mediterranean Basin, situated at the point where the vestiges of Laurasia and Gondwana meet, now includes plants that share a common origin with other plants growing very much further south. The tree heather is related to heathers that form part of the floral kingdom of the Cape of Good Hope in South Africa.

Bottom right: Peeling bark on the branches of a Cyprus arbutus (*Arbutus andrachne*). When it peels, the bark of the Pacific madrone (*Arbutus menziesii*), which grows in California, is so similar to that of the *Cyprus arbutus* that one could easily confuse them. The Pacific madrone and the *Cyprus arbutus* are 'vicariant' species, in other words species that remain very similar after the separation of their originally shared distribution area.

dentate margins, and the ability to regrow from the root-stock after fire. The plane trees that decorate village squares in Provence are a hybrid of the Eastern Mediterranean plane (*Platanus orientalis*) and the North American plane (*Platanus occidentalis*). The close relationship between these two planes explains why their hybrid offspring is fertile. Our common plane tree, named *Platanus × acerifolia* or *Platanus × hispanica* depending on which author you consult, has escaped from the roadsides where it was planted to naturalize along watercourses in the South of France, evoking the common ancestor of its parents which lived more than 65 million years ago.

Some species which for a long time lived on one or the other side of the North Atlantic following the fragmenta-

Top: Inflorescence of *Salvia leucophylla*, a Californian sage that grows in our garden. More than 50 genera of plants that grow around the Mediterranean are also found in the south-west of the United States.

Bottom: A row of plane trees along the River Vidourle at Sommières. The plane tree that adorns villages in the South of France is a hybrid, a cross between the Eastern Mediterranean plane (*Platanus orientalis*) and the North American plane (*Platanus occidentalis*). The common ancestor of these two species lived in Mediterranean regions more than 65 million years ago.

tion of Laurasia became extinct in one of these regions as a result of changes in the climate that they were unable to tolerate. Hollies and pistachio species disappeared from America, for example, at the time of the last Ice Age. Sequoias (*Sequoia sempervirens* and *Sequoiadendron giganteum*), giant conifers found today only in California, once also lived in southern Europe, at a time when the Atlantic had long separated North America and Eurasia. On the island of Lesbos, in the eastern Aegean, there is rare evidence of the history of the flora before the arrival of the mediterranean climate: in the petrified tropical forest at the western point of the island can be seen the fossils of a mixed forest, which included palms, persimmons, yews, camphor trees and conifers, among them sequoias. Petrified as a result of a volcanic eruption more than 18 million years ago, the trunks of the sequoias of Lesbos, some with a circumference of more than 15m (49ft), now lie on the ground among ballotas (*Ballota acetabulosa*), centaureas (*Centaurea spinosa*), spiny burnet (*Sarcopoterium spinosum*) and the almond-leaf pear (*Pyrus spinosa*), conjuring up the history, in part shared, of the floras of California and the Mediterranean Basin, which are today often used together in Mediterranean gardens.

• *A dead pigeon and a salt crust*

The extraordinary history of the colonization and dispersal of plants from the Mediterranean region was extended when new lands arose, such as the Canary Island archipelago. Of volcanic origin, the Canary Islands progressively arose from the Atlantic as a result of massive accumulations of lava from many successive eruptions, starting about 20 million years ago in the case of the oldest islands of the archipelago. The island of Tenerife is the gigantic cone of a volcano of which the base is on the sea floor at a depth of about 3000m (9842ft), while its summit rises to more than 3700m (12,139ft), its peak, Teide, frequently being snow-covered. Lanzarote and Fuerteventura lie closer to Africa, about 100km (62 miles) from the African coast. To colonize these newly emerged volcanic lands, the plants that grow today on the Canaries must have crossed the sea, some spread by the wind. This is the case for plants in the Asteraceae family, for example the vicariant groundsel species with succulent stems that we see today both in southern Morocco (*Kleinia anteuphorbium*) and in the Canary Islands (*Kleinia neriifolia*): their seeds have long silky plumes that allow them to be lifted by rising air currents and to float on the wind.

Yet how did species with heavier seeds manage to reach the Canaries? In the 1850s, Charles Darwin was fascinated by the puzzle of how plants colonized islands in the Atlantic. With his colleagues he carried out experiments on the possible different means by which plants were transported. Darwin wondered, for instance, whether the dead bodies of birds floating on the sea might be vectors for seeds; he succeeded in germinating seeds of leguminous plants that had spent 30 days in the crop of a dead pigeon floating on the sea. He considered too that migrating birds might carry seeds. Thus he meticulously collected dried mud from the feet of migratory birds and counted the seeds he found there under a dissecting microscope, cleaning them in order to germinate them. He also immersed the seeds of many species in seawater to find out how long they would remain viable: of the 87 species tested, 64 were able to germinate after immersion for about a month, while some seeds were still viable after spending 137 days immersed in seawater. Darwin then tried to make seeds float. He experimented first with decorticated seeds, then with whole fruits or seed capsules, and finally with fruiting bodies still attached to dry twigs which served as rafts for the seeds, in order to estimate the potential distance travelled depending on the sea currents and the number of days spent afloat. He concluded that 14 per cent of the species he tested had seeds capable of crossing 1487km (924 miles) of sea while still retaining their viability.

Thanks to Darwin's experiments, we have a better understanding of how the pines, arbutuses, laurels, cistuses,

Top: *Euphorbia officinarum* near Mirleft in southern Morocco. The plants that grow in the Canary Islands today must have crossed the sea to colonize the volcanic land that started to emerge 20 million years ago.

Bottom: The Canary Island euphorbia (*Euphorbia canariensis*) is related to the cactiform euphorbias that grow on the coasts of Morocco.

buglosses, aeoniums, sideritis, euphorbias and lavenders that abound in the Canary Islands were able to cross the sea to colonize these new volcanic lands progressively. The colonization of the Canaries by Mediterranean plants also demonstrates to what extent the long-distance migration of seeds, whether by land, sea or air, may have played a vital part in the enrichment of the Mediterranean flora by plants of extremely diverse origins. The mixing was faster in some epochs, while in others it may have been more limited as a result of the topography of the emerging lands and changes in the climate, leading to a distinct diversification of genera and species, for instance between the Eastern and Western Mediterranean.

About 5.6 million years ago, in a period when the climate around the Mediterranean was changing to a dry tropical one, the Mediterranean underwent the most serious ecological crisis in its history. This episode, called the 'Messinian Salinity Crisis', on the one hand increased the possibilities of plant exchanges between some islands but on the other strengthened the ecological distance

Top: The succulent-stemmed groundsel (*Kleinia anteuphorbium*) in the Macaronesian steppes of southern Morocco. Groundsel seeds are equipped with long silky plumes, allowing them to be lifted by rising thermals and float on the wind over long distances.

Bottom: *Cistus chinamadensis* is one of the four pink-flowered cistus species that grow in the Canary Islands. Cistus seeds are contained in capsules which can float on the sea for a long time, making it possible for the plant to spread to very distant places thanks to sea currents.

developing between the Western and Eastern regions of the Mediterranean. The diverging genetic evolution of the populations of myrtle (*Myrtus communis*) in the Eastern and Western Mediterranean goes back to this period. The Messinian Crisis originated from the tectonic upheaval in the Gibraltar Arc caused by the collision of the African and Eurasian plates, which sealed off the Mediterranean from the Atlantic Ocean. Deprived of its main source of water, the Mediterranean rapidly became desiccated by evaporation within a few thousand years. The resulting shallows aided the exchange of plants by forming bridges linking some islands, as for example Mallorca and Menorca, and opening up new migration possibilities between Sardina, Corsica and Liguria, between Sicily, Lampedusa, Malta and Tunisia, and between the Aegean islands and the coast of Turkey. At Gibraltar plants moved freely between Africa and Europe. The abyssal plains, lying at a depth of more than 1500m (4921ft) below the initial sea level, were transformed into a mosaic of highly saline lagoons surrounded by immense deserts of salt where temperatures were in the region of 35–40°C (90–104°F), preventing any plant migrations: the monumental salt crust that formed during the Messinian Crisis, now buried beneath the sediments at the bottom of the Mediterranean, is in some places as much as 3000m (9842ft) thick. The great rivers such as the Rhone, the Ebro and the Nile hollowed out gigantic canyons at the bottom of the desiccated Mediterranean, and the narrow channel which gradually opened at the Gibraltar threshold let in salt water which continually evaporated, adding to the thickness of the layer of salt.

The end of the Messinian Crisis, 5.3 million years ago, was extremely brutal: under the pressure of the Atlantic, the Gibraltar threshold collapsed abruptly and the ocean poured in through the breach to fill the Mediterranean once again in the space of only a few years. The level of this new sea rose extraordinarily fast, by as much as 10m (33ft) per day, isolating the islands again and re-establishing the movement of seeds by sea between areas that had been cut off by the salt for a few hundred millions of years.

At last the Mediterranean climate appeared, between 3.2 and 2.6 million years ago – in other words, shortly before the beginning of the great migratory ebbs and flows that accompanied the glaciation cycles of the Quaternary. The last traces of the tropical flora shrank and then mostly disappeared, although some rare species took refuge in very restricted areas where a few exceptional relics have survived to our own day: examples of these are *Myrica faya* and *Rhododendron ponticum*, which are found in a few valleys with a mild and damp microclimate on the west-facing side of the Serra de Monchique in southern Portugal. Many of the plants that make up the Mediterranean flora today, such as cistuses, phlomises, sages, thymes, *Sideritis*, *Teucrium* and lavenders, which existed before the Mediterranean climate appeared, diversified rapidly during this period. New Mediterranean species mingled with the pre-existing flora, originating in a tropical climate but nevertheless adapted to the climate that was becoming Mediterranean, for example phillyreas, pistachio species, arbutuses, oaks and olives. The climate fluctuations that followed over the next 2.6 million years – almost to our own times – led to an exceptional mixing of plants of which the history, age and geographical origins differed enormously, making the Mediterranean Basin a biogeographical crossroads that is unique in the world. Today, in the composite flora that forms the landscapes of the Mediterranean, some of which we see in Mediterranean gardens, there are plants with very different origins, for example:

• Those coming from the semi-arid steppes of Central Asia, where the summers are burning and the winters cold and dry, including artemisias, ephedras and saltbushes (*Salsola* and *Suaeda*), as well as pistachio species, cypresses (*Cupressus sempervirens*), Judas trees (*Cercis siliquastrum*), storax (*Styrax officinalis*), hackberries (*Celtis australis*) and ashes (*Fraxinus*);

• Species influenced by the flora of Africa, including grasses such as *Hyparrhenia hirta*, helichrysums (*Helichrysum italicum*, *H. stoechas*), heathers (*Erica arborea*, *E. manipuliflora*, *E. multiflora*), retamas (*Retama sphaerocarpa*, *R. monosperma*), capers (*Capparis spinosa*), dwarf palm (*Chamaerops humilis*) olive, phillyreas (*Phillyrea angustifolia*, *P. latifolia*), the jasmine of the garrigues (*Jasminum fruticans*), tamarisks (*Tamarix gallica*), oleanders (*Nerium oleander*) and the Cretan palm (*Phoenix theophrasti*);

• Those influenced by the northern flora of Eurasia, usually present in cooler and more humid regions, such as maples (*Acer monspessulanum*, *A. sempervirens*), whitebeams and service trees (*Sorbus aria*, *S. domestica*), plane tree (*Platanus orientalis*), walnut (*Juglans regia*), hazel (*Corylus avellana*), elm (*Ulmus glabra*), sweet chestnut (*Castanea sativa*), yew (*Taxus baccata*), holly (*Ilex aquifolium*) and beech (*Fagus sylvatica*);

• And finally those influenced by plants that grew on the shores of the Tethys Ocean, the ancestor of the Mediterranean, in an epoch when the climate was tropical, before they diversified markedly when the Mediterranean climate appeared, such as cistuses, germanders, *Sideritis*, thymes, lavenders, rosemaries and sages, as well as arbutuses (*Arbutus unedo*, *A. andrachne*), myrtle (*Myrtus communis*), carob (*Ceratonia siliqua*), junipers (*Juniperus oxycedrus*, *J. phoenicea* subsp. *turbinata*, *J. thurifera*), bay (*Laurus nobilis*), pines, cedars and numerous oaks.

Top left: The leaf size and fruit colour of myrtles (*Myrtus communis*) can vary in the wild depending on the population. The diverging genetic evolution between the myrtle populations of the Eastern and Western Mediterranean dates back to their separation during the Messinian Salinity Crisis more than five million years ago.

Top right: *Rhododendron ponticum* is a relict species which took refuge in enclosed valleys offering a mild and damp climate on the west-facing side of the Serra de Monchique in southern Portugal.

Bottom: The laurel-leaved cistus has curious red flower buds, quite different from those of other cistus species. Cistuses existed before the appearance of the mediterranean climate and became diversified over a period between 3.2 and 2.6 million years ago.

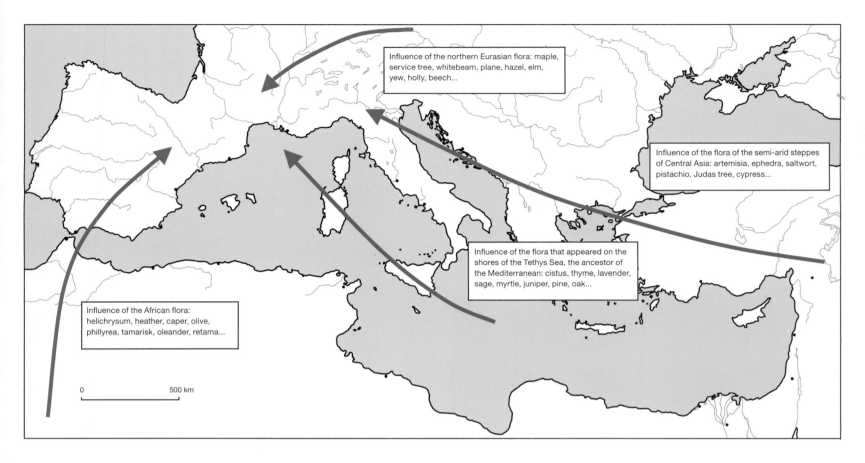

Influence of the northern Eurasian flora: maple, service tree, whitebeam, plane, hazel, elm, yew, holly, beech...

Influence of the flora of the semi-arid steppes of Central Asia: artemisia, ephedra, saltwort, pistachio, Judas tree, cypress...

Influence of the flora that appeared on the shores of the Tethys Sea, the ancestor of the Mediterranean: cistus, thyme, lavender, sage, myrtle, juniper, pine, oak...

Influence of the African flora: helichrysum, heather, caper, olive, phillyrea, tamarisk, oleander, retama...

0   500 km

The Mediterranean Basin is a biogeographical crossroads unique in the world, the result of an exceptional mixing of plants with different historical and geographical origins.

• *Migrations of the past and future*

The biologists Antoine Kremer and Rémy Petit have studied the migration of deciduous oaks that followed the end of the last Ice Age approximately 11,700 years ago. Their study of fossilized pollen together with genetic analysis showed that deciduous oaks, such as the sessile oak (*Quercus petraea*) and the pedunculate oak (*Q. robur*), which today form the forests of northern France and England, took refuge in the South during the last Ice Age, in the three Mediterranean peninsulas and in North Africa. Confined for several tens of million years to southern regions where the milder climate allowed them to survive, these oaks subsequently and very rapidly moved back towards the north once the post-glacial climate began to warm up. Progressing with an average speed of 500m (1640ft) per year, they were able to conquer the whole of Europe and about 6000 years ago stabilized in their current distribution zone which extends northwards to Scotland and Sweden. It is probable that humans, who used acorns as food, contributed to the speed with which these oaks spread. In southern Europe, however, the apogee of deciduous oak forests was brief; human influences

on the environment, which began just as the oaks were regaining their distribution area, caused a profound modification of the structure of forests. Deciduous oaks such as downy oak (*Quercus pubescens*) were to a large extent replaced by holm oak and its associated sclerophyllous plants which were able to exploit disturbances to the environment thanks to their ability to regrow from the rootstock after felling or fire had occurred. A two-way migration was thus taking place on the northern side of the Mediterranean, profoundly transforming its landscapes: a north-south post-glacial movement and an east-west movement with the spreading of new species that accompanied the expansion of agriculture and the opening of the environment by humans.

According to the botanist Pierre Quézel, one of the founders of modern biogeography, post-glacial migrations did not cease a few thousand years after the end of the last Ice Age; on the contrary, the expansion dynamics of some species are still perceptible today, although they are often inextricably linked to human activities, which as far as plants are concerned are simply another means of dispersal.

The Judas tree (*Cercis siliquastrum*) lights up garrigues in springtime with its bright pink flowers, for instance in the valleys of the Gardiole massif south of Montpellier or in the calanques of Marseilles near the Col de la Gineste. One of the oldest surviving Judas trees in France is the specimen planted by Richer de Belleval in the Jardin des Plantes in Montpellier at the end of the 16th century, its twisted trunk one of the attractions of the garden today. Brought back from the Near East at the time of the Crusades, the Judas tree quickly escaped from gardens. It is now naturalized in Italy, the South of France and Spain, in all of which countries it shows a strong tendency to spread. The Judas tree self-seeds easily on roadsides and in garrigues. Above the village of Sauve, in the Gard Department, it abounds near the 'sea of rocks', an extraordinary chaotic landscape consisting of enormous blocks of limestone shaped by erosion, where it grows among spindle (*Euonymus europaeus*), turpentine trees (*Pistacia terebinthus*) and maples (*Acer monspessulanum*). Present in the South of France during the interglacial periods, the Judas tree is now regaining the distribution area that it occupied in an earlier epoch thanks to the way it is spread by humans.

The Atlas cedar (*Cedrus atlantica*) is also a pioneer species that colonizes abandoned garrigues. Although it self-seeds in the calanques of Marseilles, for example at the bottom of the Mestrallet valley to the north of Mount Puget or above Lake Salagou near the village of Octon, it is principally at higher altitudes that it shows a strong tendency to spread from the zones where it was planted as a forest tree. It can be seen on the edges of the Navacelles cirque, on the plateau of Larzac or near Lourmarin in the Luberon. The magnificent cedar forest of the Luberon, which today

The human influence on the landscape, which began when oaks had only just finished recovering their distribution area after the last Ice Age, led to a profound modification of the forest structure: deciduous oaks were largely replaced by evergreen oaks such as the holm oak and the kermes oak, seen here on the edge of the Lassithi plateau in Crete.

The flattened seedpods of the Judas tree (*Cercis siliquastrum*) contain seeds that germinate easily, allowing the plant to escape from gardens. The Judas tree grew in the South of France during the interglacial periods, and is today reclaiming the distribution area that it occupied in the distant past thanks to the way it is spread by humans.

Opposite top: Alfa grass (*Stipa tenacissima*) is common in the hot, dry regions of southern Spain and the Maghreb and has already become locally naturalized in the South of France, where by the end of the century ecological conditions may become similar to those prevailing in the Maghreb today.

Opposite bottom left: Native to the South and East of the Mediterranean Basin, *Cenchrus setaceus*, a grass that loves heat and drought, colonizes the banks of a canal in Sète. The future recomposition of plant communities and the appearance of empty niches may favour the presence of garden escapes in human-influenced environments such as wild or semi-wild environments.

covers 250 hectares (618 acres), calls to mind the landscapes around Chefchaouen in the Rif Mountains of northern Morocco. It is the result of the natural spreading of the cedar from a small population sown in 1861 between the hamlets of Lourmarin and Bonnieux. Today the tranquil atmosphere of the century-old cedars attracts crowds of visitors and a discovery trail has been opened with educational panels giving information on the history of the cedar, the ancestors of which were one of the major components of the flora of the South of France in the period preceding the last ice ages.

Past and current migrations risk becoming more extensive in the future. With climate change, the landscapes of the South of France will probably experience a new migratory movement of substantial dimensions. The ecologists Jean-Luc Dupouey and Vincent Badeau have studied the potential impact of climate change on forest trees in France, among them beech and holm oak. They have, for example, modelled the potential distribution area of holm oak based on a characterization of its climatic niche and various different scenarios of a warming climate. In an optimistic scenario corresponding to an increase in temperature of about 2.5°C (4.5°F) by the end of the century, the potential distribution area of holm oak and its accompanying group of sclerophyllous shrubs may move northwards through France so dramatically that it may even pass the limit represented by the Loire.

Other researchers plan ahead for the behaviour of forest trees in the north of France by investigating stocks that are genetically adapted to drought and heat. Brigitte Musch, head of the Conservatoire génétique des arbres forestiers at the Office national des forêts, is selecting beeches from the forests of the South of France, for example the forest of the Sainte-Baume, between Marseilles and Aix-en-Provence, to genetically enrich the local populations of beeches in north-east France. Through the Giono programme, which concerns the assisted migration of southern populations, more than 7000 beech trees grown from rootstock from the South of France have been planted in the forest of Verdun. The aim of this programme is to establish gene flows that are crosses between the original northern populations and the new populations from the south, in order to give the forests a chance to evolve and adapt to future climate conditions.

In the South of France, modelling the future distribution of plants gives rise to questions. The landscapes of old abandoned garrigues will certainly no longer follow the model of sclerophyllous plants evolving towards a forest stage dominated by holm oaks, as is familiar to us today. The flora of the South of France might be enriched by new species coming from the south or east of the Mediterranean Basin. According to Pierre Quézel, by the end of the century the ecological conditions of the South of France could come to resemble the conditions current in the northern Maghreb. The foreseeable increase in the expansion of species such as the Aleppo pine and the partial replacement of sclerophyllous plants by a thermophile scrub comparable to that of the Mediterranean coast of North Africa, including for example the dwarf palm (*Chamaerops humilis*), alfa grass (*Stipa tenacissima*) and esparto grass (*Lygeum spartum*), risk increasing the frequency of fires, which themselves speed up the transformation of the structure of the vegetation. This predicted migration of plant species linked to changes in the climate will no doubt lead to a revision of our ideas about which plants are indigenous and which are not in any given time

Left: Tree medick (*Medicago arborea*), native to the Eastern Mediterranean, now has a distribution area that has spread to the Italian littoral and the South of France.

and place. The changing composition of floral associations and the appearance of unoccupied niches could also lead to an increased presence in the wild of garden escapees such as *Phlomis fruticosa*, *Medicago arborea*, *Ampelodesmos mauritanicus*, *Stipa tenacissima* and *Cenchrus setaceus*, marking a new phase in the evolution of the cultural landscapes of the Mediterranean.

• *The challenges of the future*

In our garden, inspired by both the garrigue and the cultural landscapes of the Mediterranean, we have sought to evoke the history of the intermixing of plants around the Mediterranean. The trees that we have planted include holm oak, azarole (*Crataegus azarolus*), Montpellier maple and Aleppo pine, which come from our nearby garrigues,

but also cypresses, introduced by the Romans, Judas trees, introduced by the Crusaders, almonds, introduced by the Greeks, and black mulberry (*Morus nigra*), introduced into the South of France to feed silkworms in the days of the silk industry. Our small collection of olives includes varieties brought by the Phoenicians, the Romans and the Arabs, and the varieties of pomegranate that we grow originated in Spain, Tunisia and Turkey. Our collection of cistuses includes species and hybrids from all over the Mediterranean, brought back over the course of the centuries by botanical explorers: some of our cistuses come from the Akamas peninsula in Cyprus, some are from the Sierra de Ronda in southern Spain, several come from Cape Saint Vincent in the south of Portugal, and others come from the Cap des Trois Fourches in Morocco, near the Spanish enclave of Melilla.

One section of our garden, consisting of plants with a cushion or ball shape, is inspired by the cultural landscapes of Crete and the Cyclades, where the vegetation has been constantly sculpted by sheep and goats ever since they were first introduced in the seventh millennium BCE. In this section, among scabiouses from Crete, anthyllises from Spain and heathers from Turkey, we grow a collection of sages, including *Salvia pomifera*, native to Greece, *S. multicaulis*, native to Lebanon, *S. officinalis*, native to Croatia, and *S. lavandulifolia* subsp. *blancoana*, from Spain. With these Mediterranean species are sages that come from the California chaparral, which blend so well into the garrigue garden that it is often hard to recognize their different origin, such as *Salvia leucophylla* and the magnificent carpeting *S.* 'Bee's Bliss', probably a hybrid of the Sonoma sage (*S. sonomensis*) which grows to the south of San Francisco. Their silvery foliage that remains attractive in summer, their intense scents, their flowers rich in nectar, their allelopathic properties and their ability to live in poor, well-drained soil without ever needing water-

ing make these California sages first-rate plants to enrich the range of garrigue plants in Mediterranean gardens.

We have planted various arbutuses in our garden which have become some of its finest plants. We grow the common arbutus from our local garrigue (*Arbutus unedo*) which has white flowers, but also its magnificent pink-flowered form (*A. unedo* var. *rubra*). The latter grows wild in south-west Ireland, still living in the coastal zone where it took refuge during the last Ice Age. We also grow the Cyprus arbutus (*Arbutus andrachne*) and the Arizona arbutus (*A. arizonica*) which we have planted in the same bed; when I look at these two arbutuses growing side by side I try to imagine the time when the common ancestors of the arbutuses of North America and of the Mediterranean lived together on the supercontinent of Laurasia about a hundred million years ago.

One of the arbutuses that has grown fastest in our garden is the Villa Thuret arbutus (*Arbutus × thuretiana*), which is a hybrid of the *Cyprus arbutus* and the Canary Island arbutus (*A. canariensis*). This magnificent plant, without doubt the most spectacular arbutus in our garden, recalls the partly common history of the flora of the Mediterranean Basin and that of the Canary Islands. As I stroke the curious smooth red trunk of the Villa Thuret arbutus, I think of Darwin's dead pigeon and the extraordinary history of the colonizations and migrations of Mediterranean plants.

The Villa Thuret in Antibes is a botanic garden with a remarkable collection of trees native to different mediterranean-climate regions of the world. Here, for example, you may see the Pacific madrone (*Arbutus menziesii*), native to California, the Chilean wine palm (*Jubaea chilensis*) which comes from the Valparaiso area, *Leucadendron argenteum*, a magnificent member of the Proteaceae family with silky grey leaves that grows on the slopes of Table Mountain above Cape Town, and some imposing Australian eucalyptus species, such as the white-trunked

Eucalyptus dorrigoensis; the specimen planted at the Villa Thuret in 1913 is now 38m (125ft) tall. The Villa Thuret recently celebrated its 150th anniversary: the continual acclimatization of species from other parts of the world for more than a century has contributed to the identity of the cultural landscapes of the Côte d'Azur. Today the Villa Thuret's scientific team, directed by Catherine Ducatillon, is engaged in a new departure related to what is at stake today in the acclimatization of plants; as well as the investigation of newly introduced plants for their robustness, their beauty and their usefulness in gardens and green spaces, each species is also assessed for its possible risks to health, for instance as an allergen, and its potential invasiveness if it escapes from gardens into the wild. For Catherine Ducatillon, it is a question of avoiding, for example, the introduction of pyrophyte members of the Proteaceae family belonging to the genus *Hakea*, native to Australia, whose naturalization in the siliceous soils of the Les Maures or Estérel massifs could lead to management problems and increase the risk of fire.

In the same spirit, background work was recently carried out by the Filière Horticole Française to categorize plants considered to be invasive and to provide a code of conduct on invasive plants (see www.codeplantesenvahissantes.fr). These new moves offer an alternative to the sometimes inappropriate stigmatization of plants considered to be potentially invasive: the protocols allow us to arrive at a reasoned definition, without any *a priori* dogmas, of the restrictions to be recommended on the use of such plants in gardens and green spaces, based on a detailed analysis of the possible positive or negative impacts related to the use of the plants in different types of environment.

Left: The Hottentot Fig (*Carpobrotus edulis*) colonizes rocks beside the sea near Centuri, Cap Corse. In some conditions the carpets of *Carpobrotus* can create management problems or compete with rare or heritage species living in fragile ecosystems. The code of conduct on invasive plants proposes restrictions on their use based on analysis of the positive and negative impacts, in different types of environment, of plants considered to be invasive.

Bottom: The phenomenon of invasiveness does not concern only exotic plants. Here *Phragmites australis*, an indigenous reed, forms vast monospecific stretches along the banks of the River Durance.

One of the plants whose invasiveness may pose problems around the Mediterranean in coming years is the Aleppo pine (*Pinus halepensis*), which is propagated by fire. Here, fifteen months after a fire, Aleppo pines have self-seeded abundantly on the Croatian island of Brač.

Ecologists who specialize in biological invasions, for example Mark Davis in the USA, Ken Thompson in the UK and Jacques Tassin in France, show that the risks of invasion do not relate only to exotic plants. In the Mediterranean Basin many indigenous plants can also become locally invasive, such as for example the Aleppo pine (*Pinus halepensis*). Their spreading can pose management problems when a change in the disturbances to the landscape alters the evolutionary trajectory of the ecosystem. According to the geographer Charles Warren, the spatio-temporal scale that defines ideas of what is indigenous and what exotic is often arbitrary. The administrative frontiers of countries or regions bear no relation to the dynamic biogeographical distribution of plant species, and the date of 1492, generally chosen as the cut-off point for distinguishing exotic from indigenous plants, makes little sense in the context of the history over thousands of years of exchanges of plants around the Mediterranean

Basin. The landscapes of the Mediterranean are the result of an extraordinary history of mixings and migrations: what characterizes garrigues is not so much their local flora, which corresponds to a particular moment in the long history of the landscape's evolution, as their ability to undergo a permanent transformation with the arrival of new plants, linked to changes in the climate or the human footprint on the landscape. The aesthetic, cultural and ecological interactions established betwen species of different origins are the most striking characteristic of Mediterranean landscapes. Reflecting these composite landscapes, the garrigue garden can profit from the rich heritage of the migrations and introductions of plants that progressively became emblematic of the landscapes around the Mediterranean. By bringing Mediterranean plants into your garden you become part of that rich heritage, reaching out from your garrigue garden to the whole world.

Top: Almond trees naturalize on the margins of cultivated land all along the N'Fiss valley in the Atlas Mountains. The aesthetic, cultural and ecological interactions established between species of different origins form the most striking hallmark of Mediterranean landscapes.

Bottom: A landscape in full evolution near the Col de la Gineste between Marseilles and Cassis: Judas trees (*Cercis siliquastrum*) mingle with maples (*Acer monspessulanum*), kermes oaks (*Quercus coccifera*) and pines (*Pinus halepensis*). Reflecting the composite landscapes of the Mediterranean, the garrigue garden can make the most of the rich heritage offered by the migration and introduction of plant species.

A garrigue garden
open to the world,
inspired by the
complex and
fascinating history of
Mediterranean
landscapes.

A garrigue garden open to the world

In our garden, largely inspired by the ecosystems of the garrigue, plants coming from other parts of the world find a place, such as the pomegranate, the almond or the nut-bearing pistachio, all of them native to Asia, but also the pepper tree (Schinus molle) native to South America and the 'desert sage' (Leucophyllum frutescens) native to the American Southwest and northern Mexico. The list below includes an indicative selection of plants representing different botanic families and different biological types, coming from other parts of the world, all of which we have integrated into our garden. This list is intended only to provide food for thought. If we included all the species from the various mediterranean-climate regions of the world and from the transitional zones of mountains and deserts, the list would run to more than 75,000 species that could be integrated into a garden inspired by the landscapes of the Mediterranean, depending on local climate and soil conditions, the desired aesthetic and the cultural perceptions of each gardener.

Acacia karroo (Southern Africa)

Allium tuberosum (East Asia)

Aloe striatula (South Africa)

Arbutus arizonica (Mexico, South-West United States)

Artemisia tridentata (Western United States)

Atriplex canescens (South-West United States)

Brahea armata (Mexico)

Buddleja myriantha (Tibet)

Buddleja salviifolia (South Africa)

Callistemon salignus (Australia)

Ceanothus 'Concha' (California)

Choisya ternata (Mexico)

Clematis tangutica (East Asia)

Crocus sativus (probably originally from Asia Minor)

Dasylirion longissimum (Mexico)

Epilobium canum (California)

Eriogonum arborescens (California)

Fremontodendron sp. (California, Mexico)

Grevillea juniperina (Australia)

Hesperaloe parviflora (Mexico, Southern United States)

Heteromeles arbutifolia (California)

Jasminum humile (Himalayas, Kashmir)

Myrsine africana (Africa, Asia)

Olea europaea subsp. *cuspidata* (Africa)

Perovskia atriplicifolia (Afghanistan)

Pistacia chinensis (East Asia)

Pistacia vera (Iran, Afghanistan)

Prunus ilicifolia (California)

Psephellus bellus (Caucasus)

Punica granatum (from the Caucasus to Afghanistan)

Rosa moschata (probably Central Asia)

Salvia chamaedryoides (Mexico, Southern United States)

Salvia canescens var. *daghestanica* (Caucasus)

Salvia leucophylla (California)

Tanacetum densum (Anatolia)

Teucrium hircanicum (Caucasus, Iran)

Tulipa clusiana (Iran)

A DATABASE TO GUIDE PLANT SELECTION FOR A MEDITERRANEAN GARRIGUE GARDEN

To complement this book we have put together a large amount of information on growing species that are adapted to a garrigue garden in a database. To help plan different zones in a garden, the database facilitates plant selection according to technical, aesthetic and functional criteria. To access this search engine, go to the website *www.olivierfilippi.com* and click on 'Plants'.

The best way to use this search engine is to apply successive filters, putting the criteria in order of priority. For example, plants could be filtered as follows:

1. By type of use (groundcover plants, plants for hedges, plants for paved areas, plants for a gravel garden and so forth);
2. By technical criteria (height, hardiness, resistance to drought, resistance to salt spray, position, nature of the soil);
3. By criteria that relate to aesthetics or sensual pleasure (flowering season and flower colour, colour of foliage, aromatic plants, scented flowers);
4. By functional criteria (attractiveness to birds or beneficial insects, honey plants, allelopathic plants);
5. By behavioural criteria (growth rate, lifespan, self-seeding potential or potential to spread by vegetative means).

It is possible to apply all the criteria or to use only a few of them, depending on what you are looking for. As an example, here are some types of selection that could be made using the search engine:

• Groundcover plants with allelopathic properties;
• Aromatic plants that grow in a cushion or ball shape;
• Plants for a flowering steppe that self-seed easily;
• Grey-leaved plants that tolerate salt spray;
• Hedging plants that have berries attractive to birds;
• Plants that attract beneficial insects, classified by flower colour and flowering season.

Many additional searches could be useful too, depending on the particular conditions of each garden. To go a bit further, you might want to associate plants that complement one another in a particular section of the garden, selected by successive searches – for example, you could select evergreen shrubs to occupy the space in the long term (such as *Pistacia lentiscus*, *Phillyrea angustifolia* and *Myrtus communis* subsp. *tarentina*) and complement them with plants that have shorter lives but that grow fast and are able to self-seed abundantly in empty spaces (such as *Dorycnium hirsutum*, *Centranthus ruber* and *Catananche caerulea*).

In the list of plants that comes up in response to a search, each species is followed by a link giving important technical information such as the height and width of the plant, its hardiness and drought resistance, and its requirements as regards soil and situation.

BIBLIOGRAPHY

ALBERTINI, L., *Agricultures méditerranéennes. Agronomie et paysages des origines à nos jours,* Actes Sud, Arles, 2009.

ALEXANDRIAN, D. ET AL, *Le Feu dans la nature, mythes et réalités,* Garrone, B. (ed.), Écologistes de l'Euzière, Prades-le-Lez, 2004.

AUBERT, S., *'Les adaptations au froid',* in F. Hallé (ed.), Aux origines des plantes, Fayard, Paris, 2008.

BADEAU, V., DUPOUEY, J.-L., CLUZEAU, C., DRAPIER J. AND LE BAS, C., 'Modélisation et cartographie de l'aire climatique potentielle des grandes essences forestières françaises', in D. LOUSTEAU (ED.), *Rapport final du projet CARBOFOR 'Séquestration de carbone dans les grands écosystèmes forestiers de France',* INRA, 2004.

BELL, N.C. AND ALTLAND, J., 'Variety Trials: Growth, Flowering, and Cold Hardiness of Rockrose in Western Oregon', *HortTechnology,* 20 (3), 2010, pp. 652–9.

BLONDEL, J., ARONSON, J., BODIOU, J.-Y. AND BŒUF, G., *The Mediterranean Region, Biodiversity in Space and Time,* Oxford University Press, 2010.

BRETON, C. AND BERVILLÉ, A., *Histoire de l'olivier,* Quae, Versailles, 2012.

BYGRAVE, P., *Cistus: A Guide to the Collection at the Chelsea Physic Garden,* Page, R.G. (ed.), NCCPG Series, Chelsea Physic Garden Co., 2001.

CHATTO, B., *Beth Chatto's Gravel Garden*, Frances Lincoln, London, 2000.

–, *The Dry Garden*, Orion Publishing Co., London, 1993.

CHAZEL, L. AND CHAZEL, M., *Découverte naturaliste des garrigues*, Quae, Versailles, 2012.

COLLECTIF DES GARRIGUES, *Atlas des garrigues: regards croisés,* Écologistes de l'Euzière, Prades-le-Lez, 2013.

CUCHE, P., *Plantes du Midi*, Édisud, Aix-en-Provence, 2005.

DAGET, P., *Atlas d'aréologie périméditerranéenne, Naturalia Monspeliensia*, 1980.

DAVIS, M., *Invasion Biology,* Oxford University Press, 2009.

DUBOST, M., 'Pastoralisme et feux en Corse. Recherche de synthèses : pour en sortir', *Méditerranée,* t. 72, *Les Grandes Îles de la Méditerranée occidentale,* 1991, p. 33–8.

GILDEMEISTER, H., *Gardening the Mediterranean Way: Practical Solutions for Summer-dry Climates,* Thames and Hudson, London, 2004.

GORINI, C., SUC, J.-P. AND RABINEAU, M., 'Le déluge et la Crise messinienne', *in* EUZEN A., JEANDEL, C. AND MOSSERI, R. (eds.), *L'Eau à découvert*, CNRS, Paris, 2015.

JAUZEIN, P., 'Biodiversité des champs cultivés: l'enrichissement floristique', *Dossier de l'environnement de l'INRA,* 21, 2001, pp. 43–64.

JOUFFROY-BAPICOT, I., VANNIÈRE, B., IGLESIAS, V., DEBRET, M. AND DELARRAS, J.-F., '2000 Years of Grazing History and the Making of the Cretan Mountain Landscape, Greece', *PLOS ONE,* 11 (6), 2016.

KEELEY, J., BOND, W., BRADSTOCK, R., PAUSAS J. AND RUNDEL, P., *Fire in Mediterranean Ecosystems*, Cambridge University Press, 2012.

KREMER, A. AND PETIT, R., 'L'épopée des chênes européens', *La Recherche*, 342, 2001, pp. 40–3.

LAPEYRIE F., 'The Role of Ectomycorrhizal Fungi in Calcareous Soil Tolerance by 'Symbiocalcicole' Woody Plants', *Annales des sciences forestières,* 47 (6), 1990, pp. 579–89.

PASCAL, M., LORVELEC, O. AND VIGNE, J.-D., *Invasions biologiques et extinctions*, Belin, Paris, 2006.

PONS, A., 'The History of the Mediterranean Shrublands', *in* DI CASTRI, F., GOODALL D. AND SPECHT, R.(EDS.), *Mediterranean-Type Shrublands,* Elsevier, Amsterdam–New York, 1981.

PONS, A. AND SUC, J.-P., 'Les témoignages de structures de végétation méditerranéennes dans le passé antérieur à l'action de l'homme', *Naturalia Monspeliensia*, 1980.

QUÉZEL, P., *Réflexions sur l'évolution de la flore et de la végétation au Maghreb méditerranéen*, Ibis Press, Paris, 2000.

QUÉZEL, P. AND MÉDAIL, F., *Écologie et biogéographie des forêts du bassin méditerranéen*, Elsevier, Paris, 2003.

RACKHAM, O. AND MOODY, J., *The Making of the Cretan Landscape*, Manchester University Press, 1996.

SANGUIN, H., MATHAUX, C., GUIBAL, F., PRIN, Y., MANDIN, J.-P., GAUQUELIN T. AND DUPONNOIS, R., 'Ecology of Vertical Life in Harsh Environments: The Case of Mycorrhizal Symbiosis with Secular Cliff Trees (*Juniperus phoenicea* subsp. *turbinata* L.)', *Journal of Arid Environments*, 134, 2016, pp. 132–5.

SAX, D., STACHOWICZ, J. AND GAINES, S., *Species Invasion, Insights into Ecology, Evolution and Biogeography*, Sinauer Associates, Sunderland, 2005.

SUC, J.-P. AND CLAUZON, G., 'La crise de salinité messinienne, une histoire fabuleuse', *Bulletin de la Société fribourgeoise de sciences naturelles*, 85, 1996, pp. 9–23.

SUC, J.-P., Origin and Evolution of the Mediterranean Vegetation and Climate in Europe. *Nature*, 307, 1984, pp. 429–32.

TASSIN, C., *Paysages végétaux du domaine méditerranéen*, IRD éditions, Marseille, 2012.

TASSIN, J., *La Grande Invasion. Qui a peur des espèces invasives* ?, Odile Jacob, Paris, 2014.

TAYLOR, J.M., *Phlomis, The Neglected Genus*, NCCPG, 1998.

THOMPSON, J., *Plant Evolution in the Mediterranean*, Oxford University Press, 2005.

THOMPSON, K., *Where Do Camels Belong? The Story and Science of Invasive Species*, Profile Books, London, 2015.

TRABAUD, L., 'Modalités de germination des cistes et des pins méditerranéens et colonisation des sites perturbés', *Revue d'écologie* : la Terre et la Vie, 50, 1995, pp. 3–14.

TYRWHITT, M.J., *Making a Garden on a Greek Hillside*, Denise Harvey Publications, Evia, 1998.

UCKO P, AND DIMBLEY, G., *The Domestication and Exploitation of Plants and Animals*, Transaction Publishers, Chicago, 2008.

UPSON T. AND ANDREWS, S.P., *The Genus Lavandula*, Royal Botanic Gardens, Kew, 2004.

VERNET, J.-L., *L'Homme et la forêt méditerranéenne de la Préhistoire à nos jours*, Errance, Paris, 1997.

INDEX OF BOTANICAL NAMES

Page numbers in italics indicate that the names appear in the captions to the illustrations.

Fraxinus 262
Fraxinus ornus 46, *47*, 149
Fremontodendron 275

G

Galium aetnicum 124
Galium verum 163
Genista acanthoclada 33, 95, *95*, 250
Genista aetnensis 12, *163*
Genista linifolia 85
Genista lobelii 50, 204
Genista umbellata 39, 56, 66
Geranium sanguineum *75*, *165*, 166, *172*, 198
Ginandriris sisyrinchium 58, *59*
Gladiolus 133
Gladiolus italicus 29
Glaucium flavum 108, *109*, 112, 167
Globularia 10, 133
Globularia alypum 26, *41*, 60, 125
Globularia meridionalis 12, 46, *47*, 79, 229
Goniolimon speciosum 139, 178
Grevillea juniperina 275

H

Hakea 270
Helianthemum 24, 26, 79, 226, 257
Helianthemum apenninum 50, 198, *199*, 229
Helianthemum canariense 209
Helianthemum oelandicum 28
Helianthemum syriacum 29, 56, *57*
Helichrysum 10, 39, 88, 120, 130, 133, 135, 139, *139*, *141*, 163, *170*, 215, *217*
Helichrysum italicum 24, 40, 52, 53, *59*, 114, 117, 125, *138*, *174*, 262
Helichrysum italicum subsp. microphyllum 44, *174*, 201
Helichrysum litoreaum 48
Helichrysum orientale 54, 138, 139, 149, 169
Helichrysum stoechas 107, 139, 198, 262
Hellebore 10, 153
Helleborus 215, *223*
Hertia cheirifolia 139, *141*
Hesperaloe parviflora 275
Heteromeles arbutifolia 275
Hyparrhenia hirta 13, *15*, 48, 66, 101, *104*, *150*, *156*, *176*, *179*, 180, *248*, 262
Hypericum 161
Hypericum balearicum 30, 54, *54*, 97, 139, 161
Hypericum olympicum *221*, *226*, 228
Hyssopus 223, 227
Hyssopus officinalis 229

I

Iberis procumbens 58
Ilex aquifolium 262
Inula candida 112
Inula verbascifolia 109, *110*
Iris 135, 172
Iris foetidissima 48

Iris lutescens 25, 50, *172*
Iris pseudopallida *24*, *41*, 46, *75*
Iris unguicularis 167
Iris unguicularis subsp. cretensis 44

J

Jasminum fruticans 229, 262
Jasminum humile 275
Jubaea chilensis 270
Juglans regia 262
Juniperus 25, *38*, *70*, 182
Juniperus oxycedrus 25, 104, 201, *206*, 262
Juniperus phoenicea *24*, *31*, 53, 83, 86, 121, 251
Juniperus phoenicea subsp. turbinata 24, 58, 82, 104, 249, 262
Juniperus phoenicea subsp. turbinata var. turbinata 52
Juniperus thurifera 25, 60, 134, 262

K

Kleinia anteuphorbium *24*, 259, *261*
Kleinia neriifolia 259

L

Laserpitium gallicum 50
Launaea cervicornis 54
Laurus nobilis *160*, 180, 262
Lavandula 23, 67, 88, *88*, 120, 127, 139, 142, 169, 209, 214, 215, *217*, *218*, *223*, 227, 240, 262
Lavandula angustifolia 112, 198, *214*, *225*, *226*, *228*, 229, *254*
Lavandula dentata *11*, 24, 29, 56, 125, 222
Lavandula dentata var. candicans 60, *111*
Lavandula lanata 56, 220, 250
Lavandula latifolia 50, 107, 167
Lavandula maroccana 60
Lavandula multifida 56
Lavandula pedunculata 58, 126
Lavandula spica 26
Lavandula stoechas 52, 56, 82, *115*, 134, 142
Lavandula viridis 142
Lavandula × chaytorae 107, 164, 229
Lavandula × ginginsii 138, *179*
Lavandula × allardii 157, *222*
Lavandula × chaytorae 214
Lavandula × intermedia *155*, *228*, 229
Leontopodium nivale subsp. alpinum 201
Leucadendron argenteum 270
Leucanthemum graminifolium 28
Leucophyllum frutescens 275
Ligustrum vulgare 182
Limoniastrum monopetalum 138
Limonium pruinosum 178
Linaria ventricosa 60, *61*, 95
Linum arboreum 44, 104, *201*
Linum campanulatum 228
Linum flavum 25
Linum narbonense 26, 229

Linum perenne 39
Linum suffruticosum *228*
Lithodora fruticosa 56, 161
Lithodora hispidula 113
Lithodora hispidula subsp. versicolor 42
Lithodora rosmarinifolia 48
Lithospermum 25
Lomelosia 87
Lomelosia albocincta 44, *220*
Lomelosia crenata 138
Lomelosia crenata syn. Scabiosa crenata 48
Lomelosia cretica 54
Lomelosia hymettia 130, 164
Lomelosia minoana 12, *13*, 55, *67*, 121, 157, 160, *160*, *163*, 221, *255*
Lonicera 25, 50
Lonicera etrusca 182
Lonicera implexa 182
Lygeum spartum 169, *181*, 266

M

Malva sylvestris 74, 157
Mandragora 10
Mandragora officinarum 42
Marrubium 10
Marrubium cylleneum 201
Marrubium supinum 229, *269*
Marrubium vulgare 79
Matthiola sinuata 48
Medicago arborea 101, 149, *267*, 268
Medicago lupulina 146, 157
Melica ciliata 163
Melilotus officinalis 157
Mercurialis tomentosa 107
Micromeria 120, *226*
Morus nigra 268
Muscari 133
Muscari botryoides 48
Myrica faya 141, 262
Myrsine africana 275
Myrtus 10, 23, 24, 53, 67, 69, 71, *72*, 114, 117, 132, 138, 149, 182, *184–5*, 186, 222, 262
Myrtus communis *24*, 52, 95, 96, *97*, *176*, 199–200, 261, 262, *263*

N

Narcissus 20, 26, 133, 165, 178
Narcissus papyraceus 178
Narcissus tazetta 48
Nepeta 130, *218*
Nepeta argolica 198
Nepeta tuberosa 166
Nerium oleander 98, *100*, 262
Nigella 223
Nigella damascena 91, 166, 242, *243*

O

Olea europaea 23, 30, *31*, 44, 69, *72*, 102–3, 112, 149, *195*, 237, 240–1, *241*, *242*, 262

Salvia lavandulifolia subsp. *lavandulifolia* 229, 255

Salvia lavandulifolia subsp. *oxyodon* 108, *225*, 229

Salvia lavandulifolia subsp. *vellerea* 56, 108, *115*, 176, 221, 229, 250

Salvia leucophylla 258, 269, 275

Salvia multicaulis 85, 88, 229, 269

Salvia officinalis 46, *47*, 79, 114, *174*, 229, *254*, 269

Salvia pomifera 13, 44, *45*, 82, 107, *112*, *163*, 164, *232*, 269

Salvia sclarea 108, 109, *152*, 153, 161, 167, *225*, *254*

Salvia sonomensis 269–70, *270*

Salvia tomentosa 38, 176

Salvia verbenaca 166

Salvia viscosa 167

Santolina 24, 64, *88*, 120, 127, 130, 145, 163, 169, *170*, *188*, 215, *217*, 227

Santolina chamaecyparissus 50, *64*, *154*, 175

Santolina elegans 56

Santolina etrusca 229

Santolina insularis 164, *225*, *254*

Santolina magonica 125, *127*

Santolina neapolitana 48, *158–9*, 229

Santolina rosmarinifolia 163, 175, 229

Santolina villosa *132*, 175

Saponaria sicula 124

Sarcopoterium 13

Sarcopoterium spinosum 23, *24*, 33, 44, 95, *95*, 127, 138, 250, 259

Satureja 215, 227

Satureja montana 46, 229

Satureja spicigera 224

Satureja spinosa 201, 229

Satureja subspicata 229

Satureja thymbra 20, 44, 114, 134, 178, 250

Scabiosa atropurpurea 74, 166

Scabiosa columbaria 218

Scabiosa crenata 201

Scabiosa cretica 155

Scandix pecten-veneris 242

Schinus molle 275

Sedum 25, 135, 170, *226*

Sedum album 229

Sedum ochroleucum 50, 229

Sedum sediforme 50, *218*

Senecio cineraria 52, *138*

Senecio inaequidens 240

Sequoia sempervirens 259

Sequoiadendron giganteum 259

Sideritis 227, 262

Sideritis cypria 42, *43*, 88, *89*, 146

Sideritis syriaca 44, *45*, 107, 165, 200, *224*, 229

Sideritis trojana 221

Silene colorata 141

Silene fruticosa 42, *43*, 109, 221

Smilax aspera 25, 182

Sonchus tenerrimus 157

Sorbus aria 50, 262

Sorbus domestica 262

Spartium junceum *160*, 229

Sphaerophoria scripta 156

Stachys 172, 227

Stachys byzantina 109–10, 163, *172*, 201, *204*, *221*

Stachys candida 201

Stachys cretica 107, *112*, 146, *158–9*

Stachys euboica 198

Stachys germanica 166

Stachys glutinosa 52, 117, 125, *174*

Stachys lavandulifolia 229

Stachys olympica 201

Staehelina dubia 107

Staehelina petiolata 44, 99, 167, *201*

Stauracanthus genistoides 11, *24*, *59*, *126*

Stauracanthus spectabilis 58, 125–6

Sternbergia 164

Sternbergia lutea 165

Stipa 156

Stipa barbata 146, 167, *168*, 178

Stipa calamagrostis *47*, 169

Stipa gigantea 169

Stipa pennata 39

Stipa tenacissima 24, *57*, 266, *267*, 268

Styrax officinalis 42, 183, 262

Suaeda 262

Syncarpia glomulifera 134

T

Tamarix gallica 262

Tanacetum 227

Tanacetum cinerariifolium 24, 46

Tanacetum densum *174*, *176*, 177, 178, *178*, *226*, 229, 275

Tanacetum siculum 124

Taxus baccata 262

Tetraclinis articulata 60, 209, 255, 257

Teucrium 16, *18*, *85*, 130, 133, 138, 149, 172, 215, 222, 227, 262

Teucrium aroanium 201, 229

Teucrium aureum 26, 38, 50, *87*, 107, 198, *199*, 229

Teucrium brevifolium 127

Teucrium capitatum 52, *52*, 109, 117, 125, *174*

Teucrium capitatum subsp. *majoricum* *15*, 30

Teucrium chamaedrys 229

Teucrium cossonii 108, 165

Teucrium creticum 42

Teucrium cuneifolium 200

Teucrium divaricatum 20

Teucrium flavum 167, 169, 229

Teucrium fruticans 25, 60, 107, *121*, *138*, 155, 157, *158–9*, *160*

Teucrium hircanicum 275

Teucrium lusitanicum 250

Teucrium lusitanicum subsp. *aureiforme* 56, *108*

Teucrium luteum 108, *174*

Teucrium marum 52, *96*, 108, 117, *177*, 178, 224

Teucrium musimonum 222

Teucrium pseudochamaepitys 29

Teucrium subspinosum 54, 55

Teucrium vincentinum 10, *11*, 58

Thapsia villosa 58

Thymbra 127

Thymbra capitata 117, *118–19*, 120, *125*, 127, *176*

Thymbra capitata syn. *Coridothymus capitatus* 13, *15*, *16*, 23, *24*, 32, 44, 117, 125, 175, 178, 250, *254*

Thymus 24, 26, 30, *36*, 38, 67, 79, *84*, *85*, 127, 135, 163, 170, 175, *175*, 214, 215, *223*, 226, 262

Thymus camphoratus *11*, 58, 59, 125, 175, *177*

Thymus dolomiticus 199, 224, *254*

Thymus leucotrichus subsp. *neiceffii* 108

Thymus longicaulis 39, *39*, 46, 79, 120

Thymus longiflorus 56, *57*

Thymus mastichina 56, 175

Thymus munbyanus subsp. *ciliatus* 108, 178

Thymus praecox *176*

Thymus saturejoides 25, 60, *61*, 175

Thymus thracicus 175

Thymus vulgaris 28, 50, 115, 175, 197, 198, 199, 221, 228, *254*

Thymus zygis 56

Trachelium caeruleum 56

Tragopogon 165

Tragopogon porrifolius 74, *76*, 157

Trifolium fragiferum 146

Triticum aestivum 245

Tuberaria guttata 142

Tulipa 26, 164

Tulipa clusiana *232*, 275

U

Ulmus glabra 262

Urospermum dalechampii 74, 157

V

Verbascum 113, *113*, *188*, 215, *223*, 227

Verbascum spinosum 44

Verbascum thapsus 46

Viburnum 24, 25, 67, 70, 72, *72*, 87, 138, 146, 149, 169, 180, 182

Viburnum tinus 50, 99

Vitex 149

Vitex agnus-castus 32, 138, *163*, 182, *240*

W

Widdringtonia wallichii 257

Z

Zelkova abelicea 122

Ziziphus lotus 196

LIST OF COMMON NAMES

A plant may often have a variety of common names in English which can lead to confusion. For this reason it is worth learning the botanical names of the plants that you grow as these enable all gardeners everywhere in the world, regardless of their country or native language, to be certain that they are referring to the same plant.

African olive	*Olea europaea* subsp. *cuspidata*	Clary sage	*Salvia sclarea*	Hackberry	*Celtis australis*
Albaida broom	*Anthyllis cytisoides*	Common plane	*Platanus × hispanica*	Hawksbeard	*Crepis sancta*
Aleppo pine	*Pinus halepensis*	Coolatai grass	*Hyparrhenia hirta*	Hazel	*Corylus avellana*
Alfa grass	*Stipa tenacissima*	Corncockle	*Agrostemma githago*	Heather	*Erica* spp.
Algerian iris	*Iris unguicularis*	Cornflower	*Cyanus segetum*	Holly	*Ilex aquifolium*
Almond	*Prunus dulcis*	Cretan palm	*Phoenix theophrasti*	Holm oak	*Quercus ilex*
Almond-leaf pear	*Pyrus spinosa*	Cretan scabious	*Lomelosia minoana*	Honeysuckle	*Lonicera* spp.
Ash	*Fraxinus ornus*	Cretan thyme	*Thymbra capitata*	Hop-hornbeam	*Ostrya carpinifolia*
Aspic lavender	*Lavandula latifolia*	Cupid's dart	*Catananche caerulea*	Horehound	*Marrubium vulgare*
Azarole	*Crataegus azarolus*	Cypress pine	*Callitris columellaris*	Horned poppy	*Glaucium flavum*
				Hottentot fig	*Carpobrotus edulis*
Balearic buckthorn	*Rhamnus ludovici-salvatoris*	Dalmatic iris	*Iris pseudopallida*		
Black medick	*Medicago lupulina*	Desert sage	*Leucophyllum frutescens*	Jerusalem sage	*Phlomis fruticosa*
Barbary thuya	*Tetraclinis articulata*	Dittany of Crete	*Origanum dictamnus*	Jove's beard	*Anthyllis barba-jovis*
Bay	*Laurus nobilis*	Downy oak	*Quercus pubescens*	Judas tree	*Cercis siliquastrum*
Beech	*Fagus sylvatica*	Dubrovnik centaury	*Centaurea ragusina*	Jujube	*Ziziphus lotus*
Blackthorn	*Prunus spinosa*	Dwarf iris	*Iris lutescens*		
Bladder-senna	*Colutea arborescens*	Dwarf palm	*Chamaerops humilis*	Kermes oak	*Quercus coccifera*
Bloody cranesbill	*Geranium sanguineum*			King Juba's euphorbia	*Euphorbia regis-jubae*
Bramble	*Rubus ulmifolius*	Eastern Mediterranean plane	*Platanus orientalis*		
Bridal veil broom	*Retama monosperma*			Lady's bedstraw	*Galium verum*
Broom	*Genista, Retama, Spartium* and *Stauracanthus* spp.	Edelweiss	*Leontopodium nivale* subsp. *alpinum*	Lamb's ears	*Stachys byzantina*
Buck's horn plantain	*Plantago coronopus* subsp. *humilis*	Elm	*Ulmus glabra*	Lentisk	*Pistacia lentiscus*
Buckthorn	*Rhamnus* spp.	Esparto grass	*Lygeum spartum*	Love-in-a-mist	*Nigella damascena*
Butcher's broom	*Ruscus aculeatus*	Etna bedstraw	*Galium aetnicum*		
Butterfly lavender	*Lavandula stoechas*	Evergreen oak	*Quercus rotundifolia*	Mallow	*Malva sylvestris*
Box, Boxwood	*Buxus*			Mandrake	*Mandragora officinarum*
		Fennel	*Foeniculum vulgare*	Mock privet	*Phillyrea latifolia*
Calabria pine	*Pinus brutia*	Field maple	*Acer campestre*	Montpellier maple	*Acer monspessulanum*
Candytuft	*Iberis procumbens*	Flax	*Linum* spp.	Mount Etna broom	*Genista aetnensis*
Cape cedar	*Widdringtonia walichii*	French lavender	*Lavandula dentata*	Mount Hymettus scabious	*Lomelosia hymettia*
Caper	*Capparis spinosa*				
Carob	*Ceratonia siliqua*	Germander	*Teucrium* spp.	Mulberry	*Morus nigra*
Chamomile	*Chamaemelum mixtum*	Giant cane	*Arundo donax*	Mullein	*Verbascum* spp.
Chaste tree	*Vitex agnus-castus*	Giant fennel	*Ferula communis*		
Chicory	*Cichorium*	Golden oak	*Quercus alnifolia*	Narrow-leaved mock privet	*Phillyrea angustifolia*
Chilean wine palm	*Jubaea chilensis*	Grape hyacinth	*Muscari* spp.	North American plane	*Platanus occidentalis*
Christ's thorn	*Paliurus spina-christi*	Gromwells	*Lithodora* spp.		
		Groundsels	*Kleinia* spp.		

Oleander	*Nerium oleander*	Snapdragons	*Antirrhinum* spp.	
Oleaster	*Olea europaea* subsp. *europaea* var. *sylvestris*	Spanish lavender	*Lavandula stoechas*	
		Spindle	*Euonymus europaeus*	
Olive	*Olea*	Spiny burnet	*Sarcopoterium spinosum*	
Othonopsis	*Hertia cheirifolia*	Spiny euphorbia	*Euphorbia spinosa*	
Oxeye chamomile	*Cota tinctoria*	St John's wort	*Hypericum* spp.	
		Sticky fleabane	*Dittrichia viscosa*	
Pacific madrone	*Arbutus menziesii*	Stinking bean trefoil	*Anagyris foetida*	
Palestine oak	*Quercus cocccifera* subsp. *calliprinos*			
		Stocks	*Matthiola sinuata*	
Pedunculate oak	*Quercus robur*	Sumac	*Rhus coriaria*	
Peony	*Paeonia clusii*	Sweet chestnut	*Castanea sativa*	
Pepper tree	*Schinus molle*	Syrian maple	*Acer obtusifolium*	
Pheasant's eye	*Adonis annua*			
Pinks	*Dianthus* spp.	Thrift	*Armeria pungens*	
Pitch trefoil	*Bituminaria bituminosa*	Tree euphorbia	*Euphorbia dendroides*	
Pomegranate	*Punica granatum*	Tree heather	*Erica arborea*	
Poppy	*Papaver*	Tree houseleek	*Aeonium arboreum*	
Portuguese oak	*Quercus faginea*	Tree medick	*Medicago arborea*	
Privet	*Ligustrum vulgare*	Turpentine tree	*Pistacia terebinthus*	
Pyrethrum	*Tanacetum cinerariifolium*			
		Valerian	*Centranthus* spp.	
Redoul	*Coriaria myrtifolia*	Viper's bugloss	*Echium vulgare*	
Rue	*Ruta* spp.	Viburnum	*Viburnum tinus*	
Saffron crocus	*Crocus sativus*	Walnut	*Juglans regia*	
Sage	*Salvia* spp.	Whitebeam	*Sorbus aria*	
Sarsaparilla	*Smilax aspera*	Wild carrot	*Daucus carota*	
Salsify	*Tragopogon porrifolius*	Wild madder	*Rubia peregrina*	
Saltbush	*Salsola, Suaeda*	Wild olive	*Olea europaea* subsp. *sylvestris*	
Samphire	*Crithmum maritimum*			
Savory	*Satureja* spp.	Woolly lavender	*Lavandula lanata*	
Sea holly	*Eryngium maritimum*			
Sea orach	*Atriplex halimus*	Yarrow	*Achillea nobilis*	
Service tree	*Sorbus domestica*	Yew	*Taxus baccata*	
Sessile oak	*Quercus petraea*			
Scabiouses	*Scabiosa* and *Lomelosia* spp.			
Shepherd's needle	*Scandix pecten-veneris*			
Shrubby Hare's-ear	*Bupleurum* spp.			
Smoke tree	*Cotinus coggygria*			
Smooth golden fleece	*Urospermum dalechampii*			

INDEX OF PLACES AND PEOPLE

Page numbers in italics refer to places mentioned in the captions to the illustrations.